OTHER BOOKS BY THE AUTHOR

Universal Reality: The New Theory of Everything

Understanding Time: What it is and How it Works

Consciousness Explained: It's True Nature Finally Revealed

Realization: Experiencing Reality in a World of Illusion

Reality: A Sweeping New Vision of the Unity of Existence, Physical Reality, Information, Consciousness, Mind and Time

UNIFYING RELATIVITY & QUANTUM THEORY

The Revolutionary
New Universe

Edgar L. Owen

Copyright © 2016 by Edgar L. Owen

All rights reserved under International and Pan-American Copyright Conventions

First Edition, Version 1.0, October 20, 2016

Library of Congress Cataloging-in-Publication Data

Owen, Edgar L.
Unifying Relativity & Quantum Theory: The Revolutionary New Universe / Edgar L. Owen – first ed.
p. cm.
Includes biographical references.

ISBN-13: 978-1539493549 (Edgar L. Owen)

ISBN-10: 1539493547 (Pbk.)

EdgarLOwen.info

CreateSpace Independent Publishing Platform

Printed in the United States of America

To my secret muse

PREFACE

This book presents a revolutionary new computational theory that convincingly unifies relativity and quantum theory. It's a non-mathematical presentation of the little known fundamental principles of both theories and how understanding them in the context of a computational universe leads to their unification. The theory is simple, elegant and easy for anyone with a basic science background to understand. This book has been adapted from Universal Reality, which is a complete new Theory of Everything including the unification of relativity and quantum theory (Owen, 2016).

This is a completely new approach to reality and the fundament nature of the universe based firmly in logic and established science that can't be found anywhere else. Anyone interested in the fundamental nature of reality and the universe should find this book a compelling and entertaining read. It convincing clarifies the nature of both relativity and quantum theory and their seemingly paradoxical natures from an entirely new perspective on the universe. By discovering the fundamental computational principles that underlie them both it reveals an entirely new understanding of the universe in which they are naturally unified.

This book assumes a general knowledge of modern physics and computer science, at least at the popular level, and some familiarity with the great perennial issues of philosophy will be helpful. But all that's really required is the desire to explore the deepest mysteries of reality objectively with an open mind.

This book was written primarily in an effort to clarify and further develop my own understanding of reality, but hopefully its publication will make it accessible to others as well and generate intelligent criticisms and suggestions for improvement. I personally believe it's the best, most accurate theory of the unification of relativity and quantum theory that has so far been discovered, but reality itself is always full of mysteries and surprises and is always the final arbiter of truth.

To the extent this book is an accurate description of reality it's not something I have created, rather it's reality itself revealing itself to someone who has hopefully been able to observe and study it without projecting too much of his own personal programming and prejudices onto it. Reality is continuously revealing itself to all of us in all its awesome glory, and I believe anyone willing to observe it carefully and objectively will be able to personally verify and experience the truth of most of what this book contains.

I would like to thank everyone who has helped make this book possible and encouraged me while writing it. Thanks to all of you for putting up with my unusual hermetic life style. And a special thank you to all my wild visitors, including the occasional human, and to the beauty and profundity of nature, which always inspires me with meaning and joy. Thanks to reality itself for continuously revealing itself in all its glory to those who will only look with opened eyes, and thanks most of all to my secret muse. Thank you, thank you. Thank you all!

And finally thanks to all those thinkers, scholars, scientists and visionaries throughout history without whose heroic efforts, genius and cumulative hard work this book could not have been written.

The author welcomes all comments and questions and can be contacted at Edgar@EdgarLOwen.com.

CONTENTS

INTRODUCTION 1
OUTLINE OF UNIVERSAL REALITY 6
COMPUTATIONAL REALITY 12
    THE COMPUTATIONAL UNIVERSE 12
    EXISTENCE & THE QUANTUM VACUUM 13
    COMPUTATIONAL STRUCTURE 20
    PROGRAMS & DATA 23
    CONSISTENCY & COMPLETENESS 25
    SUPER CONSISTENCY 28
    IMPLICATIONS 29
FUNDAMENTAL PRINCIPLES 31
    THE STc PRINCIPLE 31
    THE CONSERVATION OF PARTICLE COMPONENTS 32
    THE MEv PRINCIPLE 33
    THE METc PRINCIPLE 36
    IMPLICATIONS OF THE METc PRINCIPLE 38
    THE SOURCE OF THE METc PRINCIPLE 43
    EXCLUSION PRINCIPLES 43
    THE PRINCIPLE OF CHOICE 45
    SUMMARY 46
THE COMPLETE FINE-TUNING 47
    INTRODUCTION 47
    BRIEF REVIEW OF THE QUANTUM VACUUM 50
    COMPUTATIONAL SYSTEMS 51
    COMPUTATIONAL SPACE 53
    LOGICAL OPERATORS 56
    DATA TYPES 57
    ELEMENTAL DATA OF REALITY 59
    PARTICLE COMPONENTS 62
    FUNDAMENTAL CONSTANTS 66
    THE PLANCK CONSTANT 68
    ZERO-POINT ENERGY 68
    PARTICLE MASSES 70
    FOUR FORCES 73
    FOUR DIMENSIONS 75
    DEEP SYMMETRY 77
    THE REAL MICROCOSM & MACROCOSM 78
    THE ELEMENTAL PROGRAM 81
    HIGHER-LEVEL LANGUAGES 83
UNDERSTANDING TIME 86
    THE NATURE OF TIME 86

- TIME AND RELATIVITY 88
- THE STc PRINCIPLE 89
- TIME & GRAVITATION 92
- THE VELOCITY FABRIC OF SPACETIME.. 94
- ACTUAL VERSUS OBSERVATIONAL TIME DILATION 97
- TWO KINDS OF TIME 98
- SOME THOUGHT EXPERIMENTS... 101
- THE PRESENT MOMENT 104
- P-TIME... 105
- THE ARROW OF TIME 108
- CONFIRMING A PRESENT MOMENT 109
- THE HYPERSPHERICAL UNIVERSE 110
- INFLATION & THE HUBBLE EXPANSION 113
- SEEING ALL 4 DIMENSIONS 116
- SINGULARITIES IN TIME 117
- SPACE TRAVEL 118
- TIME TRAVEL.. 118
- ENTROPY & TIME 121
- CONCLUSION 123

COMPUTING RELATIVITY 125
- MASS VIBRATIONS & GRAVITATION 125
- RETHINKING SPACETIME 129
- HAPPENING & THE P-TIME PROCESSOR 132
- THE UNIVERSAL REFERENCE BACKGROUND 133
- NEWTON'S BUCKET & MACH'S PRINCIPLE... 136
- DIMENSIONAL DRIFT 138
- UNIFYING RELATIVITY & QUANTUM THEORY 142
- PROCESSOR CYCLES & THE STc PRINCIPLE 144
- PROCESSOR CYCLES & QUANTUM RANDOMNESS 145
- FOUR COMPUTATIONAL LEVELS... 147
- THE EQUIVALENCE OF MASS-ENERGY & SPACE...................................... 151
- A NEW MODEL OF GRAVITATION 154
- GRAVITATIONAL ATTRACTION 156
- THE CLOCK POSTULATE 161
- MASS VIBRATIONS & THE HIGGS FIELD... 164
- THE INCREASE OF MASS WITH VELOCITY 165
- DIMENSIONAL SPACETIME 168
- OBSERVER FRAMES 169
- A NEW DARK MATTER THEORY 173

RELATIVITY & ELECTROMAGNETISM 176
- ELECTROMAGNETISM 176
- THE ELECTROMAGNETIC FIELD 180
- THE HELICAL FIELD MODEL... 181
- PHOTONS 188
- ELECTRICITY.. 190
- THE STRONG & WEAK FORCES... 192

COMPUTING QUANTUM REALITY 195
EVENTS & THE ENTANGLEMENT NETWORK 195
THE SOURCE OF QUANTUM INDETERMINACY 197
WAVEFUNCTIONS & DECOHERENCE EVENTS 198
QUANTUM GRAVITY 204
A VISUAL MODEL OF THE UNIVERSE 206
RESOLVING THE SPIN ENTANGLEMENT PARADOX 209
FIELDS & VIRTUAL DECOHERENCE 212
DIMENSIONAL FRAGMENTS 215
THE NATURE OF PROBABILITY FUNCTIONS 220
THE SOURCE OF QUANTUM RANDOMNESS 222
RESOLVING QUANTUM PARADOX 224
WAVEFUNCTIONS 226
QUANTUM TUNNELING 227
HALF-LIVES 228
BOUND PARTICLES 230
MASS-ENERGY STRUCTURES 235

EPILOGUE - TESTING THE THEORY 237

NOTES 240
1. DERIVING RELATIVITY FROM THE STc Principle 240
2. THE BLOCK TIME DELUSION 243

BIBLIOGRAPHY 246

INTRODUCTION

This book explains how the revolutionary new theory of Universal Reality unifies general relativity and quantum theory. Universal Reality is solidly based in modern science but has discovered a completely new computational *interpretation* of the universe based on little known fundamental principles underlying the nature of spacetime and mass-energy.

1. The observable universe is a computational structure consisting of programs computing the data that makes up the universe. In this view the elementary particles that compose the universe are themselves composed of their particle components, which in turn are entirely data, and thus the observable universe consists only of data in a continual process of recomputation at its most elemental level.
2. Spacetime is not a fixed preexisting container for events but is computed by quantum events in the form of the dimensional relationships among resulting particles produced by the conservation of mass-energy. Thus the observable universe consists of a universal network of particle entanglements produced by quantum events. This entanglement network contains both the mass-energy and dimensional relationships among all the particles in the universe in the form of a single unified data structure.
3. Everything in the universe is continually moving through combined space and time at the speed of light, c. Everything is continually moving through *time* at c unless it has some spatial velocity in which case its velocity through time is reduced accordingly. This is the principle that underlies relativity.
4. All forms of mass and energy are different forms of spatial velocity. Now gravitational fields become fields of intrinsic spatial velocity and gravitational relativistic effects are also due to the fact that spatial velocity reduces velocity through time. All of relativity reduces to this single little known principle. This principle also immediately explains why all forms of mass and energy are conserved since they are simply conversions of equivalent amounts of spatial velocity of different forms.
5. Because the universe is computational, there is a processor that continually computes its happening. The manner in which this processor actually computes the universe can explain both relativity and the dimensional indeterminacy and randomness of quantum phenomena as aspects of a single computational process.

6. The universe itself can now be modeled as a single uniform field of speed of light velocity. Every point in the universe has an intrinsic c spacetime velocity. The presence of intrinsic spatial velocity corresponds to the presence of gravitational fields and reduces the velocity of time at that point.
7. Now all the particles in the universe are essentially little detached crystallizations of the spacetime background able to move relative to it. This enables them to have linear spatial velocities relative to the fabric of space. But they also 'feel' the intrinsic spatial velocities of any fields in space. Their velocities in time are thus reduced by both their linear spatial velocities and the intrinsic spatial velocities of the points in space they occupy. This model explains all relativistic effects as aspects of a single principle.
8. The processor of happening computes these principles by allocating a fixed number of cycles to computing combined space and time velocity for every separate coherent process. Thus when more cycles are allocated to computing spatial velocity the number available to compute velocity in time is reduced. And at the quantum level there are random oscillations between space and time velocity cycle allocation that produce the wavefunction and uncertainty descriptions of quantum theory.

These few simple fundamental principles lead directly to the unification of relativity and quantum theory in a manner completely consistent with established science though with a paradigm shift in how that science is interpreted.

The incompatibility between relativity and quantum theory, the two fundamental theories of the universe, has been the major unsolved mystery of modern science for many years. This book presents an entirely new computational theory that neatly and convincingly accomplishes their unification.

The reason for the apparent incompatibility of quantum theory and general relativity is due to their inconsistent models of spacetime. Both model spacetime as a pre-existing container in which events occur but the spacetime of quantum theory is a fixed structure unaffected by events. On the other hand the spacetime of general relativity is a dynamic structure that is curved by the presence of mass and other forms of energy.

Our theory solves this problem by demonstrating that rather than being a pre-existing container for events spacetime is actually computed by quantum events in the form of dimensional entanglements among particles in a manner consistent with both theories. In this way both particle structures and spacetime are computed together as the single unified information structure of the observable universe.

In addition this theory reveals several new fundamental principles that transform relativity from what often seems to be a very confusing and paradoxical theory to a very simple, straightforward and obvious theory that is easy for anyone to understand. This unified computational approach enables Universal Reality to discover literally scores of important new insights about reality that can't be found anywhere else.

For example it explains why everything in the universe is constantly moving at the speed of light through spacetime. This 'STc Principle' is a little known implication of relativity that scientists usually dismiss as a curiosity but which is actually of fundamental importance to relativity and also clarifies the nature of the present moment, the arrow of time, and has important implications for the cosmological geometry of the universe.

Another important new discovery is the fact that there are two kinds of time, clock time and the time of the present moment. The very fact that space travelers always meet up in the *same* present moment with *different* elapsed clock times conclusively demonstrates there are two different kinds of time. This has been one of the most controversial parts of the theory but a number of new examples and proofs are included that clinch its validity.

Though initially counterintuitive the fact of two separate kinds of time is what we actually experience as our most fundamental experience of reality itself, a present moment through which clock time flows. Not only that it's implied by relativity itself and is actually required for relativity to even make sense. Why this is true is explained fully in the chapter on Understanding Time.

The existence of two kinds of time immediately solves all sorts of important scientific and philosophical issues from the limits of time travel to the structure of the universe, and it confirms the most fundamental and obvious of all scientific observations, the undeniable existence of a present moment through which clock time flows.

The existence of two kinds of time is also fundamental to a computational theory of the universe because there must be a privileged computational space and time for the universe to be consistently computed within it. Only within such an absolute computational background can all the different relativistic clock times be consistently computed.

This model of a background computational space also immediately solves the previously unsolved problems of Newton's bucket (what rotation is relative to) and what world lines are actually relative to. All forms of motion that produce actual as opposed to observational relativistic effects are clearly with respect to the background computational space and time in which they are actually computed.

This model also reveals stunning new insights about the very nature of space and time, mass and energy that greatly simplify our concepts of the universe, and the details of how it's continually computed by the processor of happening that computes the data state of the universe.

Only by understanding the universe as a program running in the virtual computational space of the quantum vacuum does all this become clear. The quantum vacuum is the universal virtual medium of existence in which the observable universe exists. The complete fine-tuning of the quantum vacuum determines the fundamental structure of the observable universe that exists within it, and the quantum vacuum also contains the elemental program that continually computes the evolution of the universe.

Our computational approach also immediately solves a number of fundamental philosophical problems such as the natures of existence and consciousness and the status of the laws of nature which are covered in my main book on Universal Reality (Owen, 2016)

Universal Reality is the best, most comprehensive and consistent Theory of Everything the author has been able to discover. From a few simple, understandable and quite reasonable and verifiable assumptions naturally emerges a unified and complete Theory of Everything that is completely consistent with modern science, but revolutionizes its standard *interpretations*, and also opens the way towards important additional progress in our understanding of reality.

Most of Universal Reality is reasonably self-evident when we just look at what reality is actually telling us with open eyes and carefully analyze it in the context of the deep principles underlying established physical and cognitive science. What emerges are secrets that at once are incredibly profound but amazingly obvious when they are finally recognized for what they are.

A few points of the theory may be speculative and they all must be confirmed by further work but it all fits together neatly and elegantly into a unified whole of great explanatory power that incorporates all aspects of reality. And the ultimate test of true knowledge is self-consistency over maximum scope.

The search for the Theory of Everything is the ultimate quest, and it promises discovery of the ultimate treasure. We hope to make this quest as simple, clear and enjoyable as possible while we explore the deepest secrets of the universe where the greatest most wonderful mysteries of both reality and of ourselves are waiting to be discovered.

This is a brief introduction to how the theory of Universal Reality accomplishes the long awaited unification of general relativity and quantum theory. The complete theory is presented in detail in the body of the book. The author welcomes comments and questions, which may be directed to Edgar@EdgarLOwen.com.

OUTLINE OF UNIVERSAL REALITY

We begin with an outline of the core concepts of the theory of Universal Reality that explain how the observable universe emerges computationally from an underlying substrate or medium of existence called the quantum vacuum. This gives the proper context for the portion that covers the unification of relativity and quantum theory that is the subject of this book. The complete theory of Universal Reality is covered in the book of that name (Owen, 2016).

Existence is an originally formless substrate or medium within which all the individual **forms** of existence gain their individual existences. The formless substrate of existence and the forms of existence that exist within are all that exist. All the forms of existence consist only of **data,** which is the fundamental nature of all individual things that exist. Data as it's meaningful to observers is called **in*form*ation**. Data, forms, and information are different perspectives on the same thing, the actualized forms *of* existence that exist *within* the sea of existence.

All individual forms are forms *of* existence within an otherwise formless sea of existence in the same sense as waves are forms of water within an otherwise formless sea of water. Existence is the fundamental nature of the entire universe, and everything that exists is a data form *of* existence *within* the universal sea of existence.

The **quantum vacuum** is science's initial discovery of part of the nature of existence. Universal Reality considers existence to be the complete identity of the quantum vacuum. Thus existence and the quantum vacuum are identical and are all that exists.

1. **Intrinsic attributes** of existence:
    1.1. **Necessity**. Existence necessarily must exist and self-evidently does exist. And existence must exist because non-existence cannot exist. Thus existence has always existed and there is no need for a creator or creation event. (The big bang was an *actualization* event.)
    1.2. **Presence**. The presence of existence manifests as a universal current present moment within which all things exist and all events occur.
    1.3. **Happening**. Happening is the *universal processor* that continually recomputes the current data state of the universe in

the current present moment. Happening computes both mass-energy structures and dimensionality including local relativistic clock times running at different rates within the universal current present moment. Each complete computational cycle produces the next universal current present moment. Universal Reality calls this a *P-time* (present moment time) *tick*. Observers correlate P-time with their proper time (the time shown on their own comoving clocks).

1.4. **Logicality**. All the data of the observable universe is continually recomputed by the processor of happening according to logically consistent rules. Thus the universe is a logically consistent and logically complete computationally evolving data structure.

1.5. **Absoluteness**. Everything that exists is absolutely exactly as it is in the current present moment and once computed into existence could not be different in the slightest detail. This includes the complete fine-tuning which is the only one possible.

1.6. **Immanence**. Immanence is the self-manifesting immediate presence of existence in every one of its data forms. All things internally 'shine' with the immanence of their existence. The immanent self-manifestation of existence in things is their actual observable presence and being in reality.

   1.6.1. **Consciousness** is simply the manifestation of immanence within mind. The internal shining of immanence in things is not visible to the eyes but it manifests as consciousness in mind. The immanence of existence is what makes things actually *real* in the external world and makes things *conscious* in our mental simulations of the world. Universal Reality's theories of existence and consciousness can be found in the book Consciousness Explained (Owen, 2016)

2. **Data & programs**. The forms *of* existence that exist within the quantum vacuum, the sea of existence, consist entirely of data given actuality by the immanence of their existence. Both the actualized data of the observable universe and the virtual data of the complete fine-tuning exist within the quantum vacuum of existence. They are all data forms of existence within the universal sea of existence.

2.1. **Virtual data**. The fixed virtual data of *the complete fine-tuning* that determines the possible structures of the evolving actualized data of the observable universe. This virtual data is observable only through its effects on observable data.

   2.1.1. **Logical operators**. The set of elemental operators executed by the processor of happening the elemental program uses to compute a logically consistent observable universe. By analogy the 'machine language' of reality.

- 2.1.2. **Templates**. What particle components exist and which sets of particle components make valid elementary particles. The fact of 4 forces and 4 dimensions etc.
- 2.1.3. **Fundamental constants**. (What physics calls the fine-tuning) The precise values of the fundamental constants such as the speed of light, the gravitational constant and the free constants of the standard model.
- 2.1.4. **The elemental program**. The simple set of subroutines that actually compute all particle events that make up the observable universe in accordance with the rest of the complete fine-tuning.
- 2.2. **Actualized data.** *The observable universe* of science and everything in it is actualized data in the quantum vacuum. Actualized data is observable through its interactions.
  - 2.2.1. **Particle components**. All elementary particles are composed of their particle components, which are the elemental data structures of the observable universe. These are what physics calls the particle properties or quantum numbers of particles and include lepton and baryon number, the various charges including mass, spin, spatial and temporal parity etc.
  - 2.2.2. **Particle events**. Particles interact in events that conserve each of their particle components separately.
  - 2.2.3. **The entanglement network**. Because all properly defined particle components are conserved through events the particles exiting every event become *entangled* on each type of their particle components. Chains of successive particle events form a universal entanglement network encoding the entanglement relationships among all the particles in the observable universe back to the original big bang event. The data that makes up the *current slice* of the entanglement network is the complete observable universe as it exists in the current present moment. It encodes the data of the integrated mass-energy and spacetime structures of the entire observable universe as entanglement relationships among the particle components of individual particles.
  - 2.2.4. **Computational domains**. At the aggregate level the data of the entanglement network manifests overlapping hierarchical domains based on computational density, type and dimensional relationships. Observers tend to identify individual emergent level things on the basis of computational domains in their simulations of reality.
  - 2.2.5. **Emergent programs**. All emergent data structures act as emergent programs at the aggregate level because the

interactions of all their data elements are continually being recomputed. Due to the complete fine-tuning aggregate data structures manifest as independent entities in the same sense as ordinary computer programs composed entirely of structured sets of machine language operations do.

- 2.2.6. **Living programs**. Emergent programs that are purposeful and have *internal simulations* of their environments to varying degrees. We humans are running emergent programs being computed by happening.
- 2.2.7. **The universal program**. The single running program of the observable universe. All individual programs are integrated subroutines of a universal program whose interactions compute the data of the entire observable universe.
- 2.2.8. **Cosmology**. At the largest scale the dimensionality of the observable universe takes the form of a 4-dimensional hypersphere with the 3 dimensions of space its surface and past P-time its radial dimension back to the big bang at the center. The surface is the 3 dimensions of space in the current present moment and is the dimensional form of the entire observable universe.

3. **Principles**. The fundamental principles by which the observable universe is computed that determine its structure.
   - 3.1. **Differentiation Principle.** The single formless sea of existence has actualized into innumerable individual particle component forms according to the virtual data of the complete fine-tuning.
   - 3.2. **Computational Principle.** Everything in the universe is data that is simultaneously being computed by the processor of happening. Thus it manifests as a universal program made up of the interactions of innumerable individual running programs.
   - 3.3. **Particle Component Conservation Principle.** All particle components are conserved through all particle events. Thus particle components rather than elementary particles are the elemental components of reality.
   - 3.4. **Exclusion Principles.** All sets of particle components not allowed by the templates of the complete fine-tuning are excluded. This is why the particle component sets of colliding particles must split into allowed combinations in new particles. It also includes the Pauli exclusion principle that underlies all atomic and molecular structure. The interplay of conservation and exclusion determines the mass-energy structure of the observable universe.

3.5. The **STc Principle**. The total space plus time velocity of everything in the observable universe is always equal to c, the speed of light. This is the fundamental principle of relativity.

3.6. The **MEv Principle**. All forms of mass and energy are different forms of relative spatial velocity. Only if they are different forms of the same thing can they be interchangeably conserved by the conversion of one form of spatial velocity to another.

3.7. The **METc Principle**. The total mass-energy spatial velocity and time velocity of all processes is always equal to c, the speed of light. The METc Principle combines the STc and MEv principles. This principle underlies all the effects of general relativity.

3.8. The **STo Principle**. Space versus time oscillations of the processor of happening are the source of all quantum indeterminacy including the zero-point energy, the Uncertainty Principle, and the wavefunction descriptions of particles as explained in the next section.

4. **Computations**. The universal program and all its individual programs are computed at the particle event level by the processor of happening.

    4.1. The processor of happening is an intrinsic aspect of the quantum vacuum so all the data of the universe always exists in and is computed simultaneously by the processor.

    4.2. All processor computations occur in the current universal present moment in a non-dimensional computational space in the same sense as computer programs define computational spaces.

    4.3. Everything is computed at the particle and particle component level, by analogy the machine language of existence.

    4.4. Each separate *coherent process* is computed by an individual *application* of the processor executing the elemental program. Thus coherently entangled particles continue to be computed as a single process and the measurement of one immediately affects the others. This explains the spin entanglement 'paradox'. There is no non-locality in computational space because it's the pre-dimensional space in which dimensionality is computed.

    4.5. The processor allocates a fixed number of cycles per P-time tick to compute velocity in time and velocity in space so that the total velocity through space and time of all processes is always equal to c, the speed of light. This is the computational source of the STc Principle and all general relativity effects.

    4.6. At the quantum scale there is a continual random oscillation between processor cycles allocated to compute velocity in space versus velocity in time. This random oscillation is the source of

all quantum indeterminacy (zero-point energy, the Uncertainty Principle, and wavefunctions) as the dimensionality of space versus time is conflated at the quantum scale.
4.7. The processor of happening computes all aspects of mass-energy structures including their dimensional relationships. The dimensional entanglement relationships among observed events are what observers interpret as a physical spacetime. Thus spacetime isn't a preexisting container *for* events but is computed *by* quantum events.
4.8. Because dimensionality is computed by events as the dimensional relationships of particles the spacetime that emerges is compatible with both quantum theory and general relativity.

COMPUTATIONAL REALITY

THE COMPUTATIONAL UNIVERSE

There is a single elegant and parsimonious model of the universe that is consistent with and emerges naturally from Universal Reality's concept of the quantum vacuum as the universal medium or substrate of existence (Owen, 2016). This is a completely new *interpretation* of science that is entirely consistent with science. There are a number of key components to this theory:

9. We define the universe as the totality of everything that has reality, the totality of everything that exists and has existence.
10. The universe is a computational system. It consists entirely of information or data in a continual process of recomputation. The data that is computed is the total unified *mass-energy* and *spacetime* structures of the observable universe.
11. Thus the universe can be considered a running program. The entire universe is a single universal program that continually recomputes its evolving data state.
12. This single universal program can be understood in terms of innumerable individual programs that interactively compute all the evolving details of the universe.
13. The happening of existence is the processor that executes the universal program and all its individual programs.
14. Happening defines a single universal computational space within which the universal program runs and all the computations of the universe occur. This is a non-dimensional computational space in the same sense as computer programs define non-dimensional computational spaces.
15. All actual spacetime dimensionality is computed within this computational space and is relative to it.
16. The universal program that computes the current data state of the entire universe runs in the current universal present moment common to all processes.
17. All local clock rates are computed within the common universal present moment as a function of the amount of spatial velocity present so the total space and time velocity is always equal to c, the speed of light.

18. The combined computational space and universal present moment in which everything is actually computed defines an absolute pre-dimensional reference background with respect to which actual spacetime dimensionality is relative.
19. Absolute rotation and actual world lines are relative to this computational space in which they are computed. All *actual relativistic effects* are with respect to the computational space in which they are computed and are those that are persistent and agreed by all observers.
20. On the other hand *observational relativistic effects* are due to relative motion between observers. They are observer dependent and cease as soon as the relative motion stops.
21. The data that makes up the universe is of two types, the *observable* data that makes up the observable universe and the *virtual* (non-observable) data that determines and computes the allowable structures of the observable data. The virtual data is observable only through its effect on the observable data.
22. The virtual data of the laws of nature is an essential real component of the universe and must exist somewhere. There is only one possible location that virtual data could exist known to science and that's the quantum vacuum. Thus the quantum vacuum is the locus of the virtual data of the complete fine-tuning.

EXISTENCE & THE QUANTUM VACUUM

The idea of a universal medium or substrate of existence as a common universal active ingredient of all things that exist is already nascent in science's concept of the quantum vacuum. In modern quantum theory the quantum vacuum is a virtual realm that fills all space and from which all actual particles emerge (Wikipedia, Quantum vacuum). Thus it's reasonable to assume that the quantum vacuum supports the existence of actual particles as well, since if it didn't exist actual particles could never emerge into reality.

Quantum theory implies the quantum vacuum is a universal substrate to the existence of all actualized elementary particles. Thus real actualized particles continue to exist within the quantum vacuum once they appear and its continuing presence is necessary to support their existence. So the quantum vacuum is very similar to our notion of existence as a universal substrate to the being of all things that exist and it's reasonable to identify the quantum vacuum with the substrate of existence.

In this view science has begun to discover a little of the nature of the substrate of existence in the quantum vacuum and we merely take this to its logical conclusion. In this view the quantum vacuum is the locus of both the actualized data of the observable universe and the virtual non-observable data of the laws of nature that determine the forms and computations of the observable data.

Thus the observable universe, the elemental program that computes it, and all the virtual data templates necessary to produce the observable universe reside within the quantum vacuum, which is the actively happening substrate of existence. Together we define all the virtual data of the quantum vacuum as the *complete fine-tuning*. See the eponymous chapter for a full discussion.

So our concept of existence is really just a new *interpretation* of the already widely accepted theory of the quantum vacuum. In our theory existence and the quantum vacuum are different names for the same universal substrate or medium of existence. They are identical and we use the terms synonymously as appropriate.

The quantum vacuum is experimentally confirmed by the Casimir effect (Wikipedia, Casimir effect), and is the basis of Hawking's accepted theory of evaporating black holes (Wikipedia, Hawking radiation). Thus identifying it with our notion of existence is a simple natural step that lends considerable weight to the notion of existence as a universal substrate of being. In our theory the quantum vacuum and the substrate of existence are different names for the same thing.

EVIDENCE REALITY IS COMPUTATIONAL

A computational model is by far the most reasonable and fruitful approach to reality. The computational model of Universal Reality is both internally consistent and consistent with science and the scientific method. This may initially seem counter intuitive but there all sorts of convincing reasons supporting it.

There is overwhelming evidence that everything in the universe is its information or data only and that the observable universe is a computational system:

1. To be comprehensible, which it self-evidently is, reality must be a logically consistent structure. To be logical and to continually happen it must be computable. To be computable it must consist of data because only data is computable. Therefore the content of the observable universe must consist only of programs computing data.
2. The laws of science which best describe reality are themselves logico-mathematical information forms. Why would the equations of science be the best description of reality if reality itself didn't also consist of similar information structures? This explains the so-called "unreasonable effectiveness of mathematics" in describing the universe (Wigner, 1960).
3. By recognizing that reality is a logico-mathematical structure the laws of nature immediately assume their natural place as an intrinsic part of reality. No longer do they somehow stand outside a physical world while mysteriously controlling it. A physical model of the universe is unable to explain where the laws of nature reside or what their status is (Penrose, 2005).
4. Physical mechanisms to produce effects become unnecessary in a purely computational world. It's enough to have a consistent logico-mathematical program that computes them in accordance with experimental evidence.
5. When everything that mind adds to our perception of reality is recognized and subtracted all that remains of reality is a computational data structure. This is explained in detail below and can actually be verified by carefully analyzed direct experience.
6. We know that our internal simulation of reality exists as neurochemical data in the circuits of our brain. Yet this world appears perfectly real to us. If our cognitive model of reality consists only of data and seems completely real then it's reasonable to assume that the actual external world could also consist only of data. How else could it be so effectively modeled as data in our brains if it weren't data itself?
7. This view of reality is tightly consistent with the other insights of Universal Reality, which are cross-consistent with modern science. Total consistency across maximum scope is the test of validity, truth and knowledge (Owen, 2016).
8. This view of reality leads to simple elegant solutions of many of the perennial problems of science and the nature of reality and

leads directly to many new insights. Specifically it leads to a clear understanding of the nature of consciousness and also enables a new understanding of spacetime that conceptually unifies quantum theory and general relativity and resolves the paradoxical nature of the quantum world (Owen, 2016).
9. These insights complete the progress of science itself in reducing everything to data by revealing how both mass-energy and spacetime, the last remaining bastions of physicality, can be reduced to data as explained in Universal Reality (Owen, 2016).
10. Viewing the universe as running programs computing its data changes nothing about the universe which continues exactly as before. It merely completes the finer and finer analysis of all things including us into their most elemental units. It's simply a new way of looking at what already exists in which even the elementary particles themselves consist entirely of data while everything around us remains the same. Reality remained exactly the same when everything was reduced to its elementary particles, and it continues to remain the same when those particles are further reduced to their data.

Thus there are many convincing reasons to believe that everything in the universe consists only of its data and that the apparent physicality of things is an illusory interpretation produced by our minds. All the apparently material things of the world around us are our experiences and interpretations of various types of information forms in our mental simulations of reality and by extension in the interpretations of science based on these human simulations of reality.

First, a computational universe immediately solves the vexing problem of how nonmaterial laws of nature could possibly control a material universe they were not a part of. This is a problem that was intractable in the traditional materialistic view of science (Penrose, 2005). However, if the universe and the laws that govern it are respectively actualized and virtual types of information then it's natural that both would be part of a single computational universe. The laws of nature are simply the programmatic structure of the elemental program that continually recomputes the information state of the universe.

Thus the laws of nature, being forms of information in a reality consisting only of data, are an integral part of nature as real as the data forms that encode actual things, and thus are as real as the things of the world. The laws of nature don't stand apart from nature in some mysterious metaphysical realm while controlling it as traditional science mistakenly assumes. That the laws of nature find a natural place in our computational model of reality is strong evidence for its validity.

Second, it's quite clear that our experience of a seemingly physical universe, and everything in it, actually consists only of information in the neural circuits of our brains. While there is certainly a real universe external to our brains, the seemingly physical universe we experience our existence within is without any doubt an information construct in our brains. So if just information in our brains can produce such a completely convincing illusion of a material universe, why couldn't the actual universe external to our brains also consist only of information?

That would immediately explain why neural computations within our brain's model of reality could enable us to function so effectively within actual external reality. How could our internal mental simulation of the universe so accurately map the actual workings of the universe if the universe itself wasn't also an information structure?

Third, all the laws of science consist only of mathematical equations imbedded in a logical framework, in other words they consist only of information. How could information structures accurately describe the universe if the universe itself didn't also consist of information structures? This immediately solves the mystery of why mathematics works so incredibly well to describe the universe. Of course mathematics and logic would naturally provide the best description of a universe that was itself a logico-mathematical information structure.

Fourth, when we carefully analyze seemingly material things in our minds we find that they actually consist only of the information of what they are, and this is true of everything without exception. They all consist only of their information, the combined information of their colors, textures, forms, structures, chemical compositions and whatever else makes them up. These are all just different forms of information that in combination are interpreted by our brains as material objects. Our brain tells us these combinations of information forms make material objects but even that interpretation is just more information.

This is also confirmed by the design of robotic control programs and pattern recognition (Wikipedia, Pattern recognition). In robots able to operate effectively within complex environments internal models of themselves within their environments must be laboriously constructed and continually updated from streams of raw data input. That data is then converted into simulations of purposeful action within the model, which are in turn tested, valuated and used to control appropriate motor activities. Internally it's all based on internal data models of the robots within their environments that work only due to the data model's logical consistency with the actual data structure of external reality. All living organisms including us operate on identical principles though in much more complex systems.

The information that makes up even a simple physical object, not to mention that of a living being, is not a simple data string like the name or description of an object. It's an incredibly complex hierarchy of forms and multiple hierarchies of subprograms within subprograms, and their ongoing computational interactions and relationships with other forms and programs both internal and external. Think of the hierarchies of total information content of anything down through all its individual systems to its individual cells to the detail of every one of its elementary particles and their interactions, and that is the complete information structure that makes up that thing, and actually is that thing. These are the total running programs that things actually are.

Fifth, even modern science now has now reduced the entire materiality of the universe to just mass-energy, and spacetime. However, in the chapter on Fundamental Principles it's shown how spacetime reduces to the information of dimensional relationships, and mass-energy reduces to the information of relative motion. So even the universe envisioned by modern science naturally reduces to pure abstract information.

Sixth, accepting a universe consisting only of information doesn't change the universe that we experience around us in the least. It still appears exactly as it did before, as a material universe. The only difference is that we now realize that its seeming physicality is an interpretation of its information structure produced in our minds, and that the underlying data structure of the seemingly physical world we live in is its actual fundamental structure.

Thus it's reasonable to conclude that the data structure of the universe is its actual fundamental nature and its seeming materiality is an illusion produced by our mind as it combines all the sensory information of things into the semblance of physicality.

Thus in our theory all the programs of things that make up the universe without exception consist only of their data in a continual process of recomputation. These programs have existence because they run in the substrate of existence, and thus they become the real actual things of the world, but the fundamental nature of all these things is information given being by its presence within the substrate of existence of the observable universe.

Thus at the most fundamental level the things that make up the universe are not material or physical entities, they are simply different information forms that arise in an originally formless sea of existence, as water waves, ripples and currents are different forms of water that arise in an originally formless ocean of water. And since the things of the universe are not physical they have no individual self-substances that make them different things; the only difference between things is the differences in their information forms, the different data that distinguishes them one from another.

Information takes innumerable different forms but the fundamental nature of all the data that makes up the universe is the same; it all consists of abstract data forms that are computationally evolving in a common non-material medium of existence. The only substance of all information forms is existence itself, just as the only substance of all forms of water is water no matter how their different forms may vary.

All things in the universe consist only of information given actuality by existing in the universal medium of existence. It is their common existence, rather than any material substance, that makes them all real things. They become real things by appearing in the virtual medium of the quantum vacuum, just as water waves become real by appearing in water.

Thus all the seemingly physicality of the things of the world is actually interpretations of their information forms in our minds. The apparently physical world in which we seem to exist is our mind's internal simulation of an actual external reality consisting only of the information states of running programs. To this extent the physical world is completely an illusion, though certainly a very convincing illusion.

This is equally true even at the perceptual level. It's just a matter of becoming aware of what we actually see when we observe things. If we really take our perceptions of things apart into their individual components we find their every component reduces to the information of what it is, and that's all we experience because everything without exception is ultimately perceived in terms of its information. Only information is observable. There simply isn't any way to perceive anything except in terms of its information. Perception is information input and sensory information input is perception.

So everything is actually just its information or its data, but this is not data in the usual sense of data on a printout or even in a computer. The medium of this data, the data of actual things, is not marks on paper or electronic bits in a storage device. The data of reality exists in the medium of existence, and that's what makes it real and actual. It makes the data of existence into real actual living things. Its existence gives it immanence and being, it makes whatever form the data has into the real actual thing that has that form.

And since existence continually happens that data is continually recomputed and takes on a life of its own in interaction with the life of all the other data forms that make up the universe.

COMPUTATIONAL STRUCTURE

The following overview of the basic computational structure of the observable universe explains how elemental quantum computations generate an entanglement network among all elementary particles. This entanglement network encodes and actually is the unified data structure of all aspects of the observable universe. It includes the data of all mass-energy structures and all their dimensional relationships we interpret as spacetime.

The internal logico-mathematical consistency of the dimensional relationships of mass-energy structures is what observers interpret as dimensional spacetime in their simulations of reality. All aspects of the observable universe, including spacetime, are computed by the elemental program from elemental particle interactions.

1. The observable universe consists of all actualized (non-virtual) observable data. The actualized data of the observable universe consists of particles, particle components, and their relationships. Particle relationships consist of the entanglements among particles on their particle components generated by particle events and interactions.
2. Particle components, such as spin, mass, energy and the charges are taken as the actual elemental data components of reality, which in valid combinations create elementary particles. This is reasonable since particle components are conserved through all

particle interactions even when particles transform into other particles.
3. Every particle interaction creates entanglements among its particles on each of the particle component types involved in the interaction. The additive sum of all properly defined particle components is always conserved.
4. Together all the particle entanglements of the observable universe form an entanglement network that computationally connects all the particles in the universe and establishes relationships among their particle components.
5. The logico-mathematical consistency among all the dimensional data of the particle components is what we reify and interpret as an encompassing spacetime.
6. The entanglement network can be visualized as particle event nodes connected by the data trajectories of individual particles. It incorporates all the actualized particles in the universe and all their events.
7. The entanglement network extends back to the original event(s) of the big bang and represents the entire evolutionary *history* of the observable universe at the particle component data level.
8. Thus the entanglement network is the integrated mass-energy and spacetime structure of the entire observable universe throughout its entire history.
9. The actually existent current observable universe is the current present moment data state of the entire entanglement network.
10. All the actualized data of the entanglement network is continually recomputed by the elemental program in the current universal present moment of existence to produce the current data state of the observable universe. This includes local relativistic clock times, which are all, computed in the current present moment.
11. At the aggregate level the entanglement network of all particle data takes the form of hierarchies of overlapping computational *domains*. Domains are areas of computational density and similarity characterized by domain boundaries of lower computational density and similarity.
12. Individual observers tend to view the universal program in terms of individual things and programs on the basis of domains. For example surfers extract waves, smelt extract tides, and oceanographers extract currents from the overlapping computational domains of a single ocean.
13. Nevertheless the universal program is a single interconnected entanglement network ultimately computed on the basis of individual particle interactions by separate applications of a single elemental program.

14. This single elemental program consists of a small set of virtual subroutines that identify and compute all possible types of particle interactions. Individual applications of this elemental program compute the evolution of the entire observable universe in the form of the entanglement network.
15. Domains and observer identified things and programs are *emergent* manifestations of particle aggregates. These emergent data structures act as independent programs due to the meaningful relationships among their particles. This is analogous to how computer programs designed to perform various high-level functions all consist of meaningful sequences of a small set of elemental machine language operations that perform the actual computations.
16. Computer programs consist of pre-written sequences of code executed sequentially. But the programs of reality consist only of the data of things in a continual state of recomputation. All the actualized data of the observable universe always exists in the current present moment in the quantum vacuum. Thus it all exists within the processor of happening where all the data of the observable universe is *simultaneously* recomputed by individual applications of the processor in the current present moment.
17. A simultaneous separate *application* of the elemental program and single processor of happening recomputes each individual particle interaction process.
18. Emergent programs and processes are meaningfully structured aggregates of elementary particle data. Emergence begins at the atomic and molecular level and extends up the hierarchy of aggregate structure to the highest classical levels and beyond. Nearly all of science other than particle physics is concerned with emergent processes and data.
19. Though emergent processes are meaningfully considered as programs the emergent laws of science only *describe* them rather than actually *computing* them since all actual computations occur only at the elemental level and are computed by separate applications of the single elemental program to each coherent particle process.
20. The biological programs of living organisms including us are all characterized by internal simulations of themselves within their environments to various extents. They are also characterized by purposeful actions based on instinctual imperatives that are part of their operational software passed from individual to individual and from generation to generation encoded in their DNA.

The quantum vacuum computes the observable universe at the particle and particle component level because it contains the actual elemental program that does the computing and is the processor that executes the program. However the observable universe itself acts like a universal program at the emergent level even though everything is actually computed at the elementary particle and particle component level.

Unlike a regular computer program the observable universe consists only of current data states rather than pre-programmed code sequences. Thus the quantum vacuum computes the interactions of current data states rather than emergent level sequential code strings.

The observable universe can't contain pre-programmed code strings because that would imply a pre-determined future, which doesn't exist. And if the universe consisted of multiple pre-programmed code strings that would imply multiple possible versions of the future that would inevitably lead to irreconcilable inconsistencies. So the present must be continually recomputed from the interactions of its current data states. This is consistent with science in which the evolution of the universe consists of the continual interactions of all its particles.

There is another important way the computational universe differs from the way ordinary computers work. In ordinary computers a single data computation occurs at every processor cycle. Though a computer can have multiple processors each processor can only perform one computation per cycle. This greatly limits the overall processing power of even supercomputers.

By contrast all the data of the observable universe exists in the quantum vacuum simultaneously, and the happening aspect of the quantum vacuum is the processor that computes reality. Thus all the data of the entire universe exists within the processor of happening and is recomputed simultaneously with every P-time tick to create the next current present moment. Thus the computational power of the quantum vacuum is limited only by its processor cycle rate, which we will see below has important implications.

PROGRAMS & DATA

Universal Reality proposes that reality consists only of information or data in the form of self-modifying programs running in a universal substrate of existence, which it identifies with the quantum vacuum. There are a number of convincing reasons to accept this model of reality and it also leads to simple and elegant solutions of many of the fundamental problems of science and philosophy.

In this model the universe is a single universal running program that continually recomputes its current state, the current state of the universe, in the present moment.

All the individual things and processes of the universe are individual subprograms running interactively within the universal program. They are all computationally consistent parts of the single universal program. These individual programs computationally arise, transform, and fade as identifiable structures in continual interaction with other programs within the universal program. This overall process is the computational evolution of the universe.

The individual programs are all the actual processes of the world from the most elemental interaction of particles through human beings to cosmological processes on the grandest scale. Individual programs are processes identified on an *ad hoc* basis on the basis of the computational domains produced by the universal program and their personal meaningfulness to individual observers.

Even though the observable universe consists only of its current data state rather than code strings it acts like a single universal program that can be understood in terms of innumerable individual programs in continual interaction with each other. We just need to keep in mind that the actual computations occur only at the particle and particle component level and all the emergent level programs of the universe are aggregate manifestations of elemental computations.

This is analogous to silicon computer programs where the actual computations occur only at the level of individual machine language instructions, but structured aggregates of instructions form meaningful emergent level programs with specific higher-level functions. In a similar manner the data that makes up the universe consists of meaningful data structures, though not code sequences, at the aggregate level. One can identify code sequences but these are not stored pre-programmed sequences but simply post facto lists of sequences that have already occurred.

The aggregate data structures are those familiar to science as the compound particle structures that make up all the emergent level structures of the universe. Because these data structures are continually interacting at the particle level they act as programs and these running programs are the programs that appear to compute the universe at the emergent level.

Thus the observable universe can be considered a universal program composed of innumerable individual programs because its elemental data structures at the particle level are meaningful in aggregate at the emergent level due to the details of the complete fine-tuning which define them.

We know from information science that only data can be computed, and data can only be meaningfully computed on the basis of exact logico-mathematical operations embodying consistent rules (Wikipedia, Data (Computing)). Sequences of logico-mathematical operations on data are called programs. Information science also tells us that if a set of axioms and logical operations is consistent then all computational results of the system will be logically consistent and logically complete, that the results of all possible computations will produce a single consistent logico-mathematical system with no contradictions.

All data forms are forms of in-*form*-ation. Thus everything in the observable universe consists of forms of information or data that arise in the common medium of existence and thus gain reality as real actual things. The continual recomputation of all data forms generates the evolution of the observable universe.

CONSISTENCY & COMPLETENESS

Once we understand that the programs of the universe are simply the ongoing changes occurring in its information it's clear these changes must be computational because the only way information can change is computationally. Therefore all the processes of the universe must be running programs, and what we call the things of the universe are all the current information states produced by these programs.

Because it's computational the programmatic information structure of the universe must be a rule based logical structure. The structure and programs of the universe must follow logical rules contained virtually in the complete fine-tuning of the quantum vacuum. Thus every information state of the universe is computed according to these logical rules from its previous information state, and every information structure of the universe is logically consistent with every other.

If the universe is a computational system then it must be logically consistent and logically complete because computations operate only on the basis of logical rules and for the universe to exist those rules must be consistent and complete. Only computations produce one information state from another on the basis of consistent logical rules.

Logically complete means that every computation of the universal program and its individual subprograms will always be able to produce some result. And logically consistent means the universal program can never compute any result that is a logical contradiction to any other computational result it has produced or can produce.

There is a straightforward proof that the universe is logically consistent and logically complete. If a computational universe was not a logically consistent and logically complete system, it would tear itself apart at the inconsistencies and pause at the incompletenesses and simply could not exist. Thus the fact of its existence demonstrates the universe is in fact logically complete and logically consistent, which is in fact what we always observe.

In addition a computational universe must be logically consistent and complete to be meaningful and amenable to knowledge. Since the universe is a massively understandable system that also confirms it is logically consistent and complete.

For the universe to be logically complete and consistent the fundamental axioms of the complete fine-tuning must themselves be a complete and consistent set of rules, because it's these rules that govern the computations of the universal program. The fundamental logical rules of the laws of nature must be a consistent set of axioms that cannot generate inconsistent results. And this is necessary for the universe they compute to exist. This is one clue into the mystery of why the actual fine-tuning of our universe is likely the only fine-tuning that can exist, because it may be the only one that produces a logically consistent and logically complete universe.

Logical consistency and completeness of course refers to the computations that produce the actual information states of the universe rather than to the reasoning of individual actors within the universe. Within this overall consistency it is quite possible for human thinking based on false or incomplete premises or invalid logic to generate inconsistencies as we see far too often.

In particular the logic of human and other beings is based on the emergent logic of things and highly simplified models of reality, which ignore most of its actual computational details. This enables humans to quickly compute reasonably accurate descriptions of small fragments of reality and base fairly effective actions upon them.

But by computing on the basis of highly simplified individual details and events the complete logical consistency of reality is lost. Typically what human reasoning does is continually remap simplistic models of reality consisting of relevant sets of individual things, events and relationships, and replace them with others as needed, with no necessity of complete consistency among them. This leads to the general fuzziness and inconsistencies of human thinking, but it enables humans to reason fairly quickly and accurately from moment to moment on the basis of heuristic mental models of changing situations.

At the core of this process is employing the logic of things to model the world in terms of individual things, properties and relationships, rather than its actual enormously complex single network of computations. These individual concepts are redefined as needed to quickly model the relevant details of a current situation. This often results in dynamic overlapping identities of individual things, such as a forest when hiking or an individual tree when cutting wood. This enables humans and other species to quickly reason on the basis of minimally pertinent information sets. But to accomplish this humans lose the complete picture of the processes of reality in terms of their inherently contradictory heuristic models of reality from moment to moment.

It's important to note that the logico-mathematical system of reality isn't subject to Gödel's Incompleteness Theorem because every state is directly computed from its prior state. Gödel's theorem applies only to human mathematical systems in which it is possible to propose a well-formed statement that cannot be proven either true or false (Hofstadter, 1980). But reality doesn't make statements and then try to prove them like mathematicians do. It just computes the next results from the previous results and this can always be done, and is always consistent and complete.

Thus, unlike human mathematical systems, reality mathematics can be and must be logically complete and logically consistent. The universe produces every information state computationally and so only produces statements (data states) whose truth is automatically proven by their existence.

SUPER CONSISTENCY

The great mystery is that this process works at all because it requires a type of *super-consistency* which allows the simplistic inherently contradictory overlapping models humans have of reality to exhibit reasonable internal self-consistency sufficient to ensure effective action in an actual world that consists of enormously more complex information systems.

This super-consistency is necessarily inherent in the virtual information structure of the complete fine-tuning and is what allows us to effectively live our lives within an enormously more complex computational universe.

The logical consistency of the universe is what makes it understandable, and makes reason, knowledge and science possible. The ability to map the logic of reality in our simulations of reality makes this possible. The universal program is a completely self-consistent logical system and thus there seem to be no intrinsic limits to understanding it.

The universe acts as a running program in the medium or substrate of existence which compute it and gives it reality and the living vitality of its happening. And the immanence of the quantum vacuum gives all the information forms that make up the universe the immanence of their self-manifestation that makes consciousness of them possible.

At the heart of the reality of the universe is happening, the fact that things change, that its data is continually recomputed in the present moment in which everything exists. Happening is the life of the universe. It's what brings everything to life and gives it reality and being and immanence and makes consciousness possible.

IMPLICATIONS

Universal Reality is a science-based theory. The universe evolves according to the computational laws of nature, however it does so in a manner that logically integrates happening, the present moment, immanence and consciousness, the fundamental aspects of existence, into a single unified system.

The data of the observable universe is computed into a universal entanglement network of all particle and particle component relationships by particle events. This is the fundamental data structure of the observable universe. It consists of the computed relationships of all the particles and particle components of the universe including their dimensional relationships. This entanglement network is the unified data structure of all mass-energy relationships including their dimensional relationships and is a single integrated structure.

Because spacetime is computed as the dimensional relationships of mass-energy structures both atomic and molecular structures and the spacetime they exist within are computed together as a single unified structure consistent with both quantum theory and general relativity and it's the computational emergence of spacetime that is the key to their unification as explained in upcoming chapters.

The universe actually is its computed data. Its current data state is the observable universe in the present moment. Thus the observable universe is not the physical or material structure it appears to be but the complete data of what we humans and other species *interpret* as a physical universe, each in our own way.

Thus the true nature of every individual thing in the universe is the complete data and programs of what it is. Everything is its running program, and its snapshot at any point in time is its instantaneous data state. This includes us as well. We are the programs of ourselves, and the data of ourselves in the continual process of recomputation by our programs. While at first this may seem counter intuitive and even crazy, we can actually confirm this through careful analysis.

Science is a wonderfully accurate and comprehensive model of reality and is extensively confirmed by experiment. Universal Reality generally accepts all experimentally confirmed science but fundamentally reinterprets it. In Universal Reality what science interprets as material things in a physical spacetime container actually consists of the integrated data structures of those things and this data evolves computationally rather than causally.

The only way anything can happen is for it to be computed, and the only way something can be computed is for it to consist of data. This view has long been implicit in science itself since science consists of the same logico-mathematical structures that underlie computation. However the clear implications of this view have been suppressed by the archaic belief that reality is material and physical.

However it's easy to demonstrate that the apparent physicality of reality is an interpretation generated by our mind's simulation of reality to help us make sense of the world. How this comes about is explored in detail in the author's book Realization, (Owen, 2016), but there is plenty of other evidence.

Science itself has progressively reduced the apparent physicality of the universe to fewer and fewer elements until currently only mass-energy and spacetime remain as the last bastions of the old material world.

However as we will discover below mass-energy and spacetime also reduce entirely to data, thus demonstrating in one more way that the fundamental nature of the universe is only data. Mass-energy reduces to the data of relative motion and spacetime emerges computationally from mass-energy interactions. This makes it clear that a computational approach to reality is quite reasonable, and by adopting this view we lose nothing essential to the workings of science.

This paradigm shift leads to significant new insights about our universe, and a completely new interpretation of the reality in which we exist. It also provides conceptual solutions to many of the important unresolved problems of science such as the nature of quantum paradox, the apparent inconsistency of quantum theory and general relativity, and the source of quantum randomness.

We will begin to explore how this all works with a discussion of fundamental principles in the next chapter, and the details of the complete fine-tuning in the subsequent one. At this point the details of how particle events and atomic and molecular structures are computed together with dimensional spacetime in a manner consistent with both quantum theory and general relativity can be clearly explained.

FUNDAMENTAL PRINCIPLES

THE STc PRINCIPLE

The STc Principle states that the combined vector velocity through space and time of everything in the universe is always equal to the speed of light, c. This is a consequence of relativity that is well known to scientists though usually viewed as a mere curiosity (Greene, 1999, 2005). However it's actually a fundamental principle with profound consequences. It means that every clock in the universe runs slower in time proportional to its velocity through space so that the combined spacetime velocity through time and space always remains equal to c.

This is one of the fundamental conservation principles of reality. It means that the total velocity of everything in the universe is always c, and that velocity is distributed between velocity in space and velocity in time. This is a critically important principle that underlies relativity as well as the fundamental nature of mass and energy.

The STc Principle is a direct consequence of a fixed number of processor cycles used to compute the happening of the universe. Processor cycles go first into computing the spatial velocity of processes. The remaining cycles compute the internal evolution of processes, which manifests as their clock time rates. In this manner the processor cycles that compute each process are distributed among computing velocity in space and velocity in time so that the vector sum of velocity through space and time is always c. This is explained in more detail in the chapter on Computing Relativity.

By definition observers don't move relative to themselves, thus all the c spacetime velocity of every observer is completely through time on its own clock. Thus c is actually the speed of clock time and light moves at that velocity through space because it has no internal processes to compute and thus no velocity through time on its own internal clock. Thus c is the speed of time on all observer's own clocks and the baseline velocity of all processes in the universe.

The particular value of c in our universe provides enough time for things to happen and enough space for things to happen in. If the speed of time was zero nothing could ever happen, and if it was infinite the whole history of the universe would be over before it began, so a viable universe requires a reasonable finite non-zero value of c, which is encoded in the complete fine-tuning.

The value of c must also be quite large, as it is, relative to typical velocities through space. If it weren't the spacetime dilations of mass-energy would produce gravitation so intense as to crush all possible material structures, and routine spacetime distortions so great as to make ordinary processes unintelligible.

THE CONSERVATION OF PARTICLE COMPONENTS

The particle components that make up elementary particles are the basic computational data elements of reality. The particle components combine in valid sets to form the elementary particles that make up everything in the observable universe. Every elementary particle is composed entirely of its unique set of particle components. The particle components are the basic constituents of the observable universe.

The total values though not numbers of all particle components are conserved through all particle computations. The subroutine that computes the conservation of particle components must exactly redistribute all the incoming particle component values in any particle interaction among the resulting particles or an interaction cannot occur.

Particle component conservation in particle interactions produces the network of entanglement relationships among elementary particles that observers interpret as mass-energy structures in spacetime. Thus the conservation subroutine computationally creates spacetime in the form of entanglement networks from the interactions of elementary particles and their particle components.

There are a few rare exceptions where currently defined particle components are conserved only in combinations. But it should be possible to define a proper set of particle components such that they are always conserved.

We can identify two general classes of particle components, those intrinsic particle components that uniquely identify *types* of particles, and dimensional particle components such as spin, mass and energy. Mass is a form of energy and total mass-energy is conserved in every particle interaction. This is key to understanding how processor cycles compute quantum events as discussed in the chapter on Quantum Reality.

THE MEv PRINCIPLE

Total mass-energy is conserved in every particle interaction. All forms of mass-energy are inter-convertible in particle events. The conservation of mass-energy is a basic principle of science.

For something like mass-energy to be conserved all its different forms must be forms of the same underlying thing. In the case of mass-energy all its forms are different forms of relative motion whose values are their spatial velocities. Thus the conservation of mass-energy is always the conversion of equivalent amounts of relative velocity from one form to another, and all forms of relative velocity are equivalent to some type of mass or energy. This MEv Principle is a fundamental principle with profound consequences.

Physics already understands some forms of energy, such as kinetic energy and heat, as relative motion. Kinetic energy is half the square of the linear velocity of a mass relative to an observer. And heat energy (temperature) is the average kinetic energy of the velocities of atoms and molecules within a substance.

The electromagnetic energy carried by photon is also a form of relative motion. It's the relative motion of its vibrational frequency, the vibrational velocity of electromagnetic waves. The amount of energy a photon carries is directly proportional to its frequency, which is the velocity of its wave cycles.

With $E=mc^2$ Einstein showed that mass is a sort of frozen energy. This is easy to understand if we model masses as forms of very fine very high frequency in place vibrations. By vibrating rapidly in place they have the same amount of motion relative to all stationary observers and thus particles have fixed rest masses.

This is compatible with the observed increase of mass as velocity approaches the speed limit of light. Near the speed of light it becomes more and more difficult for linear velocity to increase so some of the linear velocity is converted to internal vibrational velocity and this increases the mass as predicted by relativity (Wikipedia, Special relativity). It is worth noting that String Theory also models particles as rapidly vibrating strings (Susskind, 2006, p. 199), (Wikipedia, String theory).

Chemical energy is the binding energy of electrons in their orbitals around atomic nuclei. Binding energy is the energy that holds all forms of atomic and molecular matter together and is responsible for the different chemical compounds. It is due to the electromagnetic attraction between negatively charged orbital electrons and the positively charged protons of nuclei.

Different molecular configurations will have different binding energies and the release of chemical energy in explosions is due to the transition from molecules with higher binding energies to molecules with lower binding energies. An explosion converts the difference in binding energies to heat energy and the kinetic energy of rapidly expelled particles. The internal velocities of the binding energies are converted into the linear and wave velocities of emitted particles.

Binding energy is due to the rapid oscillatory velocities of electrons within atomic orbitals. Like the vibrations of mass, the relative motion is locally constrained so it manifests as fixed energies to stationary observers.

When particles combine in molecules there is actually a very minute increase in total mass over that of all the combined particles. In other words some of the binding energy is converted to, or manifests as, mass due to the orbital velocities.

Nuclear energy is similar to chemical energy but involves binding energies of the quarks and gluons in nuclei that hold protons and neutrons together. Nuclear energy is due to the strong force rather than the electromagnetic force that holds electrons in their orbits. Because there is much more total velocity converted in nuclear explosions they are enormously more powerful than chemical explosions. Nuclear explosions are transitions from nuclei with higher binding energies to nuclei with lower binding energies thus resulting in the conversion of the lost binding energy to other forms of energy. The internal velocities of the binding energies are converted into the linear and wave velocities of emitted particles.

This is confirmed by the modern theory of the quarks and gluons in nuclei (Wikipedia, Quark#Mass). The total mass of these particles is just ~1% of the total observable mass of nuclei so most of the mass of nuclei is actually due to the equivalent amount of relative velocities of their quarks and gluons.

As for potential energy it's sort of an accounting trick rather than an actual form of energy. When we say that an object in a system has potential energy what we are usually saying is that there is some energy in an external system that is blocking an equivalent energy in the system under consideration. The notion of potential energy just makes it easier to isolate systems for analysis. Potential energy is the amount of energy in a system that can be released by removal of the equivalent blocking energy. A compressed spring or a weight suspended by a wire are examples.

The potential energy of a charge due to its position in a force field, including a mass in a gravitational field, is not an actual amount of existing energy but a measure of predicted future energy that doesn't yet exist. It's the potential for some form of relative motion. This is another accounting trick because the conservation of mass-energy properly applies only to conversions of actual forms of relative motion as they occur.

Universal Reality models force fields as fields of velocity density of various forms specific to the force as explained below. Thus the velocity gradient of the field generates the linear velocity of kinetic energy and total velocity is conserved.

Gravitational energy is another kind of relative velocity. A gravitational field is very effectively modeled as an area of vibrational density in space itself produced by the vibrational motion of mass or energy. This dimensional vibration effectively produces the curved spacetime of general relativity as explained in the chapter on Computing Relativity. It's because of the spatial velocity density of fields that the clocks of objects in gravitational fields run slower in accordance with the STc Principle. This is also explained in the chapter on Understanding Time.

When we understand that all forms of mass and energy are just different forms of relative motion the reason for the conservation of energy among its different forms becomes clear. The conservation of energy is just the transformation of one form of relative motion to an equivalent amount of another form of relative motion. For example the conversion of mass to energy in a nuclear explosion is just the conversion of some of the mass vibrations of particles of fissionable material into equivalent amounts of wave frequencies and linear velocities of ejected particles.

Thus all forms of mass and energy are different forms of relative velocity computed by the elemental program of the universe. Since all forms of mass and energy are relative velocities they are fundamentally data forms rather than anything physical. They are part of the data of dimensional relationships computed by the elemental program of the quantum vacuum.

Thus Universal Reality demonstrates that both spacetime and mass-energy, the two remaining fundamental components of 'physical' reality, reduce naturally to computational data without diminishing the explanatory power of science. Universal Reality leaves the enormous explanatory power of science intact, but it gives us a completely new *interpretation* of that science.

THE METc PRINCIPLE

The STc Principle states that the vector sum of space velocity and time velocity always equals the speed of light c. But from the MEv Principle we see that mass-energy is equivalent to spatial velocity. By combining these two principles we arrive at an even more fundamental principle of the conservation of mass-energy and time. This METc Principle states that the total mass-energy spatial velocity and time velocity of everything in the universe always equals c, the speed of light.

To be precise the total spacetime velocity of everything is c, and the spatial velocity component of anything is always some form of mass-energy. Thus all forms of mass-energy slow the velocity of time because they are all forms of spatial velocity. The total vector mass-energy velocity and time velocity of all processes is always c.

Thus the fundamental fabric of spacetime is composed of spacetime velocity, which has the same c value everywhere apportioned between spatial velocity, which is equivalent to some form of mass-energy, and time velocity. This is a deep fundamental principle of reality that underlies the computational unity of mass-energy and spacetime and is key to really understanding both quantum theory and general relativity.

The METc Principle is a direct consequence of the fixed cycle rate of the quantum vacuum processor that computes the observable universe. It's a direct consequence of the fact that the observable universe is a computational structure. The processor cycle rates that compute the happening of everything in the universe are allocated between calculating the internal changes of processes and their spatial motion. Motion in space is their mass or energy and the rate of change of their internal processes is their velocity through time.

Thus we have the fundamental principle of the conserved equivalence of mass-energy and time that underlies general relativity and nearly every aspect of reality. The total relative velocity of mass-energy plus the total velocity of clock time for all processes is always equal to c, the speed of light, which is the cycle rate of the processor that computes everything in the universe.

This principle demonstrates that mass-energy and clock time are two aspects of a single fundamental entity and are inter-convertible so long as the total vector velocity of both is always equal to c.

The total spacetime velocity of all things is always equal to c. The spatial velocity manifests as either the intrinsic velocity density of a field or the linear or wave velocities of a particle or object, and the remaining velocity manifests as the velocity of time, the local clock time rate. This confirms that all forms of mass-energy are just excitations of space, and space itself is a ubiquitous field of energy as Quantum Field Theory proposes (Wikipedia, Quantum field theory).

For mass-energy to exist in an individual form it has to be packaged in a valid set of particle components. It then pops out of the virtual quantum vacuum as an actual particle so there is something able to move relative to the background energy field of space itself. Thus charges are quanta of the velocity field of empty space packaged in particles.

Mass-energy fields are velocity density gradients in space, and flat space itself is the uniform velocity density of the zero-point energy. Charged particles are velocity density fields centered on velocity concentrations packaged in valid particle component sets as particles. Particles and their fields are concentrations in particle component packages of the universal field of velocity density of space. Thus everything that exists, all particulate structures, consists of arrangements of little elemental bits of space in time. All mass-energy structures are composed of little quanta of spatial velocity packaged as particles.

Spacetime velocity is what holds the observable universe open so events have room to occur. Every point in spacetime is a c valued combination of space and time velocities. The total amount of velocity in the universe holds it open and gives it the volume that it currently has which is the volume of the observable universe in the present moment.

The velocity that creates and opens the observable universe is produced by the processor of happening that injects life into the universe. The fixed number of processor cycles in every P-time tick for each process goes first to computing velocity in space and the remainder computes velocity in time. The total fixed processor cycles always compute a total velocity of c for every process in the universe and at every point in the universe.

Thus the universe can be considered to consist entirely of energy in the form of velocity. This energy is generated and maintained by the processor of happening. The amount of velocity energy is equal to c at every point in the universe and consists either of spatial velocity or velocity in time. Spatial velocity is always equivalent to some form of mass or energy. Spatial velocity energy plus time velocity is fixed and conserved and always equals the speed of light c. However velocity in time can be converted into the energy of spatial velocity so long as their sum still totals to c. Thus space and time velocity of two aspects of the same fundamental element of velocity, or happening.

IMPLICATIONS OF THE METc PRINCIPLE

Universal Reality's METc Principle is already implicit in general relativity and certainly consistent with it. Einstein's special theory of relativity demonstrated that time and space were both aspects of a single 4-dimensional spacetime, and that mass and energy are equivalent forms of a more fundamental mass-energy. And in Einstein's theory of general relativity mass-energy tells spacetime how to curve, and spacetime curvature tells mass-energy how to move. This clearly suggests an actual equivalence of spacetime and mass-energy that has not been previously acknowledged (Wikipedia, General relativity).

Universal Reality reveals this previously unrecognized connection between mass-energy and spacetime. In Universal Reality mass-energy and spacetime are essentially two aspects of the same thing. Mass-energy is the velocity shape of spacetime, and spacetime is the distribution of mass-energy.

The entire spacetime universe is simply the direct manifestation of the distribution of all the mass-energy, all the relative spatial velocity, in the universe. Both mass-energy and spacetime are ubiquitous and coterminous, each is an aspect of the other, and together they are the single observable structure of the universe. We can say that mass-energy manifests as spacetime to have room to exist.

Even flat space is the manifestation of the presence of the zero-point energy of the quantum vacuum. The existence of the zero-point energy of the quantum vacuum creates a universe of flat space for it to have room to exist. And the value of the zero-point energy defines the flatness of spacetime in terms of the value of c, which is the fixed velocity of all processes in spacetime. It is the observational value of the combined space and time velocities of everything that exists.

All other forms of mass and energy are additional distortions or relative motions in flat spacetime; they are all local shapes of spacetime in addition to the flatness of its zero-point energy. Thus spacetime itself, with all its dilations and relative motions is the observable distribution of all the mass-energy of the universe. Spacetime and mass-energy are two aspects of the single computational nexus that is the observable universe.

Any deviation from the flatness of the zero-point energy space is some additional form of mass-energy, and any velocity relative to the benchmark flatness of the entanglement network that encodes spacetime is an equivalent additional form of energy. This includes the stretch warping of space around galaxies and galactic clusters due to the uneven expansion of space around cosmic scale mass-energy distributions and leads to a new theory of Dark Matter as explained in that section.

For all forms of mass and energy to be conserved they must all be different forms of the same underlying thing, and that can only be relative motion. Mass and energy are conserved through all transformations only because equivalent amounts of relative motion are being converted from one form to another. Since the underlying nature of all forms of mass and energy is relative motion, mass-energy must actually be relative motion; it must be an aspect of spacetime itself.

This explains how particles can pop into existence out of the flat spacetime of the quantum vacuum and vanish back into it. The modern theory of the quantum vacuum is that it's composed of virtual particles that pop in and out of existence so quickly they can't be observed (Wikipedia, Vacuum state).

The METc Principle is also compatible with a big bang in which all the particles of the universe could appear out of the nothingness of the quantum vacuum presumably reducing its zero-point energy by the amount of particle energy that was actualized. Thus spacetime can crystallize into particles around valid particle component sets in specific cases, and all particles may have originated as velocity eddies in computational space. There are several experimentally confirmed effects in which particles spontaneously appear out of spacetime at very high energies or relative velocities (Wikipedia, Unruh effect).

A completely flat entirely virtual zero-point energy universe likely existed prior to the big bang in a non-observable state. Only distortions or eddies within flat spacetime are potentially observable against the background and able to make observations as well. There must be some non-uniform mass-energy distribution, some spacetime structures, for observations to take place. Without some non-uniform structure nothing can happen and there will be no computable clock time. It would be completely uniform in a state of maximum entropy and no energy could be exchanged to make anything happen.

The total mass-energy content of the universe including the zero-point energy is the source of the hyperspherical geometry of the universe as explained in the chapter on Cosmology. This energy content of the observable universe on the largest scale is the attractive force that curves overall spacetime in on itself into a hypersphere. Since there seems to be no other possible viable geometry from the perspective of P-time this very fact itself requires the existence of a zero-point energy just as the existence of the zero-point energy requires a hyperspherical universe.

Thus what is called flat spacetime is actually the minutely curved surface of the cosmic hypersphere at the largest scale. Gravitation is equivalent to a spacetime curvature. Since curved spacetime is energy, the zero-point energy itself manifests as a very slightly curved spacetime. The cosmic hypersphere is the direct manifestation of the total energy of the universe; it's the inward attractive effect of the total mass-energy of the universe.

If mass-energy is relative motion in space then if space itself begins to vibrate in the right mode that vibration can effectively crystallize spacetime into particles that pop out of the quantum vacuum. In this view particle charges are little crystals of spacetime each of which has a particular mode of vibration, a particular crystal structure, corresponding to the type of charge.

Thus the several force charge particle components including mass nucleate around other particle components to form particles that are little crystals of spacetime. For example mass will be one sort of vibrating crystal, and electric charge another. Only valid combinations of these little crystals can combine to form actual particles. Thus actual particles will be combinations of particle component crystals that pack together in allowable combinations.

Thus particles can be thought of as *phase changes* of spacetime where specific forms of relative motion crystallize into particular types of particles. Each particle is a little standing vibration of spacetime; one of a small set of possible forms of persistent localized relative velocity.

In this view the charge particle components are the possible crystal structures that can be produced from spacetime. They are localized phase changes in spacetime that are conserved in all particle interactions. And they are centered in surrounding fields of velocity density or equivalently spacetime dilation each with a specific form resulting from the vibrational mode of the charge producing it.

Only the particle components carrying the charges of the four forces, including the mass charge of gravitation, produce spacetime dilations or the equivalent velocity density fields. Other particle components like particle identity or spacetime parity don't carry vibrational energy or produce fields.

Particles are composed of valid combinations of multiple particle components. Thus some carry charges of more than one force. For example electrons carry both mass and electric charge and are combinations of two kinds of spacetime structures each with its own vibrational mode, and each with its own surrounding spacetime velocity density dilation field.

There are limits to the crystal metaphor, but by extension the entire structure of the universe can be thought of as complex combinations of a few fundamental types of spacetime crystals and how they pack together into atoms, molecules and emergent material structures according to computational rules and interactions based on their basic crystalline structures or alternately their vibrational modes and forms.

As we recall from the previous chapter, what we interpret as spacetime is the underlying entanglement network produced by the conservation of particle components in all particle interactions. It is these conservation laws that generate the relative scales of spacetime dimensionalization produced in the resulting entanglement network. So from this computational perspective as well, spacetime and mass-energy are two aspects of a single entanglement network. Both mass-energy and spacetime are aspects of the same data structure computed by the conservation of particle components in particle interactions.

In addition to the flat spacetime of zero-point energy, there are three categories of spacetime effects corresponding to categories of mass-energy. These are the vibrational relative motion of charges of the four fundamental forces, the vibrational frequencies of photons, and the linear relative motion of charges and their fields with respect to the absolute background of the underlying entanglement network.

Thus all distortions in the flatness of spacetime are forms of mass-energy in addition to the zero-point energy. There are only four possible intrinsic forms of vibration and associated spacetime dilation corresponding to the four forces, though the actual shapes of the fields of the associated spacetime dilation fields in aggregate can be quite complex.

All spacetime dilation or velocity density fields produce gravitational effects since they curve or densify spacetime. However the fields of the electromagnetic, strong and weak forces primarily affect particles carrying the same type of charge since the fields of these charges couple to damp or reinforce. However there will be some residual effect on other particles with disparate charges due to the general gravitational effect of spacetime dilations of any form.

Thus electric charges mainly attract or repel other electric charges, but their fields do produce some gravitational effects as well. This is consistent with the Einstein field equations where the presence of any form of energy curves spacetime (Wikipedia, Einstein field equations).

Though electromagnetic velocity density fields do produce measurable gravitational effects the dilation fields of the strong and weak forces are constrained to nuclear scales and are too limited in range to produce measureable gravitational effects.

Thus the four forces of nature can be understood as four types of velocity density or equivalently spacetime dilation fields produced by the particles that carry their charges. Each of these velocity density fields, in particular those produced by mass, tilts the c balance of space and time velocities so that some of the usual c velocity through time becomes velocity through space. Thus the spacetime at every point in the field is distorted to slow time and increase distance proportional to the strength of the field.

THE SOURCE OF THE METc PRINCIPLE

The happening of the quantum vacuum is the processor that computes all processes in the universe. The processor has a fixed cycle rate that sets the combined space and time velocities of all processes to c, which is the fixed velocity of all happening in the universe. This is the fixed rate at which everything in the universe is computed.

The fixed processor cycles are used to compute both the internal changes of data states and their relative motion. The relative spatial motion is computed first and any processor cycles left over compute changes to internal data states. The computational rate of change of internal data states manifests as the clock rate of the process.

The processor cycle rate is the source of the METc Principle, which states that the combined velocity of everything is always equal to c, the speed of light. The METc Principle is the source of most relativistic effects. A universe that is computed on the basis of the METc Principle automatically incorporates general relativity as explained in the chapter on Computing Relativity.

EXCLUSION PRINCIPLES

Exclusion principles are the inverse of conservation principles. They appear to be based on templates in the complete fine-tuning of the quantum vacuum though there are likely more fundamental principles at work as yet undiscovered.

The first exclusion principle determines what sets of particle components produce valid elementary particles. Most combinations are excluded for unknown reasons. This principle is responsible for the redistribution of particle components in particle interaction events.

When multiple particles collide in the same location the total particle component set doesn't produce a valid particle. This is forbidden by the exclusion principle so the particle components must be redistributed in valid sets to create new particles, which are then ejected in different directions so the exclusion principle is satisfied. When the Exclusion Principle is violated the Exclusion subroutine makes a call on the Conservation subroutine to conserve and redistribute the particle components involved into valid particle sets. If they can't be validly redistributed no interaction can occur.

The Pauli exclusion principle states that no two fermions (matter forming particles as opposed to force carrying bosons) can occupy the same quantum state (Wikipedia, Pauli exclusion principle). This principle is responsible for the structure of atoms and molecules as it determines how electrons can fill atomic orbitals. It is largely responsible for the fact that atoms and molecules are stable and occupy dimensional volume, and that two material objects cannot be in the same place at the same time.

However the Pauli exclusion principle is more accurately just a result of the computational rules that govern the interactions of individual particles and the balance of forces as explained in the section on Bound Entanglement. This principle is responsible for much of the structure of the matter in the universe. For example it's responsible for the fact that ordinary bulk matter is stable and occupies volume since electrons can't all collapse together to the lowest orbital in atoms.

The filling of orbitals by electrons in turn determines the chemical structures of atoms and gives rise to the Periodic Table of the Elements and thus determines the emergent structure of the observable universe.

Another example is the Uncertainty Principle, which asserts a fundamental limit to the precision with which pairs of observable properties of a particle known as complementary variables, such as position and momentum, can be simultaneously known (Wikipedia, Uncertainty principle). However Universal Reality interprets this as uncertainties in the structure of spacetime rather than in particles with respect to spacetime as explained in the chapter on Quantum Reality.

The Exclusion principle is involved in all particle identities and interactions, and thus in all the interactions of particulate matter in the universe, and the computational evolution of the universe.

THE PRINCIPLE OF CHOICE

The Principle of Choice, aka the Randomness Principle, states that whenever the Conservation Principle could be satisfied in multiple ways nature has no rules to decide and so either makes the decision randomly or postpones it by encoding it in the form of a probability distribution called a dimensional fragment to be decided probabilistically by subsequent events. Subsequent events decohere the probability distributions of dimensional fragments to exact dimensional values. Thus in these cases subsequent events make the postponed decisions.

Importantly this occurs in dimensional particle component interactions where there are often ranges of possible exit velocities (speeds and directions) that conserve energy and momentum particle components. It also occurs where the novel spin orientations of two entangled particles can potentially have any possible orientation with respect to a measuring device so long as they are equal and opposite so total spin is conserved.

The conservation of most particle components results in only a single possible deterministic choice, but the conservation of dimensional particle components may necessitate probabilistic choices among a distribution of possibilities.

This principle is the source of all quantum randomness and thus of all the randomness and freedom in the universe, including ultimately the free will of living beings. It's the sole reason our universe is not completely deterministic and completely predictable to any future time. One could say it's the reason the future doesn't already exist.

The operation of this principle generates entanglement networks that incorporate constrained randomness into the structure of spacetime as it's created from the interaction of elementary particles. Thus there isn't a fixed pre-existing spacetime that particles are probabilistic with respect to, but a spacetime that itself incorporates a degree of randomness in its structure as it is built. This is explained in detail in the chapter on Computing Quantum Reality.

Understanding how this principle works resolves all the apparent paradoxical nature of quantum processes. Quantum processes are paradoxical only with respect to an assumed fixed pre-existing spacetime background that doesn't actually exist. When it's understood how spacetime emerges computationally in the form of entanglement networks generated by the Conservation subroutine the quantum world is no longer paradoxical.

SUMMARY

The computational structure of the universe is a balance between the active principles of conservation and exclusion, guided by the Principle of Stochastic choice at the quantum level and the METc Principle at the relativistic level. The computational interplay of these principles driven by the processor cycles of happening produce the entanglement network, which is the combined data structure of the observable universe of mass-energy structures and their dimensional relationships that we interpret as material objects within a physical spacetime.

All the emergent laws of physics that describe the processes and structure of the observable universe derive directly from the computational interplay of these few fundamental principles.

THE COMPLETE FINE-TUNING

INTRODUCTION

In modern physics the fine-tuning is the set of irreducible fundamental constants for which there is no known source or cause. It includes the specific values of the gravitational constant, the speed of light, and the values of the constants that determine atomic structure, the free parameters of the Standard Model (Wikipedia, Fine-tuned Universe).

It's clear that if the values of these fundamental constants varied by only a small fraction the universe as we know it and in which we were able to evolve wouldn't exist. Thus it seems these constants are somehow fine tuned to values that enable our existence, or at least the evolution of intelligent life.

The fine-tuning constants are essential so far as they go but in a computational universe there are clearly a number of other fundamental types of data necessary for our universe to exist in the form that it does that are not included in the standard fine-tuning.

Universal Reality proposes a *complete fine-tuning* that includes all the data necessary to produce and sustain our universe that are not reducible to, or derived from anything more fundamental. The complete fine-tuning is the complete set of fundamental irreducible information necessary and sufficient to compute the observable universe that actually exists.

This complete fine-tuning includes the values of the fundamental constants, but also several other types of essential information including an elemental set of programmatic routines incorporating the laws of nature that actually compute reality, the even more fundamental set of logical operators by which they operate, the basic data types of the universe, the elemental particle components, the four forces and four dimensions, and the fundamental templates that determine the structures of particles and atomic and molecular matter.

This chapter describes each of these fundamental components of the complete fine-tuning in detail. The complete fine-tuning is basically the virtual data structure of the quantum vacuum that encodes and implements the fundamental principles described in the previous chapter.

Since standard physics doesn't understand the universe as a computational system it doesn't provide a location or mechanism for its fine-tuning and relegates it along with the laws of nature to a metaphysical realm for which it has no explanation and so generally ignores. The inability of modern science to understand that the universe has to be computational for anything to happen, and that the laws of nature and the fine-tuning have to exist as virtual data to facilitate this, and that there is already an accepted place where virtual data actually exists called the quantum vacuum is truly one of the fundamental failings of modern science.

But in a computational universe it's natural that the laws of nature and the complete fine-tuning exist as real virtual data in the quantum vacuum. The quantum vacuum is the only reasonable locus for the complete fine-tuning and there isn't anywhere else that it could exist. So it's only natural to assume the complete fine-tuning and the elemental laws of nature exist as virtual data in the quantum vacuum and that the quantum vacuum also contains the processor that actually computes the universe on this basis. This fundamental model is simple, elegant, straightforward and highly explanatory.

The only difference between the virtual data of the elemental program and complete fine-tuning, and the actualized data of the elementary particles that make up the observable universe is that the observable universe evolves computationally and is directly observable through its interactions while the virtual data is fixed and observable only through its effects on the actualized data. Both the virtual and actual data of the universe exist entirely as data within the quantum vacuum of existence.

An ocean of water is a useful analogy. The ocean of water is initially formless but specific forms can arise within it. It's the innate or virtual nature of formless water that determines the type of forms and dynamic interactions of the water waves, ripples and currents that can appear within it. Likewise the complete fine-tuning is the innate nature of the quantum vacuum that determines all the specific types of actualized data and computational processes that can appear within it to produce the observable universe. Thus the complete fine-tuning of the quantum vacuum determines the specifics of the observable universe that appears within it.

Just as an ocean is observable only in terms of the actualized forms that appear within it, so the quantum vacuum is observable only in terms of the observable data and observers that appear within it.

Thus the complete fine-tuning completes the picture of all the information necessary to account for the actual universe, some of which exists as actualized emergent running programs as the observable universe, and some of which exists as the virtual nature of the substrate of existence in which they run that determines their natures and the natures of their interactions. In our theory the quantum vacuum is not just the virtual reservoir from which real particles emerge but also contains the complete fine-tuning rules that support the existence of real particles and govern their behavior.

This is fairly obvious since the fine-tuning couldn't generate particles with specific properties if it was not also the source of the rules that govern the natures of those particles. There would be no way to specify the forms of actual particles as they emerged if the quantum vacuum was not the reservoir of the templates for those forms. The complete fine-tuning has to exist as data in some form somewhere, and the quantum vacuum of existence is its only possible locus.

The complete fine-tuning determined the structure of everything that came into being at the big bang, and continues to determine the fundamental structure and evolution of the universe to this day. The basic routines that govern the information of particles and their interactions include everything necessary to compute the big bang and the entire evolution of the observable universe since the big bang.

The great mystery of the complete fine-tuning is how its precise details automatically produce all the immensely complex myriads of running programs that emerge from it. It's clear that the programs of reality compute not just data but effectively compute new programs that are naturally self-organizing and self-modifying, as we see so clearly in our own programs.

The other great mystery is to what extent the complete fine-tuning is the only one possible, and if so why is it as it is. Hidden within this question is the deepest secret of the universe, and the nature of existence itself. Universal Reality provides at least a partial answer but first things first.

BRIEF REVIEW OF THE QUANTUM VACUUM

The quantum vacuum, the medium or substrate of existence, is all that exists. It's an active computational system that computes the existence and evolution of the observable universe, which exists as an evolving data structure within it.

The data of the quantum vacuum consists of the virtual data of the complete fine-tuning and the actualized data of the observable universe. The elemental program computes the actualized data of the observable universe on the basis of the complete fine-tuning.

In particular the quantum vacuum computes the observable universe according to the fundamental principles outlined in the previous chapter that are the fundamental laws of nature of the complete fine-tuning.

The actualized data of the observable universe consists of the data of particles, particle components and their relationships. This universal structure is computed as an entanglement network of particle components that encodes the data of both particle structures and spacetime as a single unified structure.

The quantum vacuum also has several qualitative non-data aspects. These include the most fundamental aspects of reality namely existence, happening, the present moment and immanence.

Existence, which Universal Reality identifies with the quantum vacuum, is the medium or substrate within which all the data of reality exists and so gains its individual existence and reality.

Existence exists, and for existence to exist is must be present. The presence of existence manifests as the current universal present moment in which everything that exists does exist, and in which all the computations of the universe are simultaneously occurring. The current present moment is the manifestation of the presence of existence and is the only moment in which anything actually exists or happens.

Happening is the aspect of existence that brings existence to life. It's the processor that simultaneously computes the continual evolution of all the data of the observable universe within the quantum vacuum in the current present moment. Happening computes all the data states of the observable universe simultaneously because it's intrinsic to the nature of the ubiquitous quantum vacuum in which they all exist.

Existence, the quantum vacuum, is a living system in the sense it's continually self-activating with no external source or cause. Existence continually happens, and the happening of the quantum vacuum is the ubiquitous processor that executes the code that continually computes the current data state of the entire observable universe in all its individual interactions.

Immanence is the self-manifestation of existence and the essential ingredient of consciousness. The universe no longer exists just as material objects in a physical spacetime container. Everything in the universe glows with the inner light of its existence because it's real and actual, and absolutely what it is, and it has an actual here-now presence beyond just its material configuration. Everything that exists is really really there. Things exist in an immanent sense and are infused with their own living reality as part of the living reality of existence.

Immanence is the source of consciousness. Consciousness is simply opening a properly configured mind to the innate immanence of things. Consciousness is not something generated by human or other minds and shown out onto things like a spotlight. It's simply the living immanent presence of representations of things within mind. Just as seeing things depends on light coming from them so consciousness depends on the unseen glow of immanence in the existence of things. The internal glow of immanence is not something visible to the eyes but is visible to the mind as consciousness.

COMPUTATIONAL SYSTEMS

To understand the computational system that computes the universe we need to look to computer systems to see how they work because their general applicability and computational power clearly derive from their being modeled on the computational system of the universe. There are clear differences but the general design must be very similar.

When programming computer programs one declares variables and constants with label names and assigns them values. Variables can be declared as different data types; typically Booleans, integers, floating point numbers, arrays and strings. One then refers to declared variables by their label names in the code of the program where they are logically and mathematically manipulated to produce desired results.

Then either at run time or before, this code is interpreted or compiled into the sequences of machine language operators in terms of which the program actually computes. The machine language operators perform elemental logical, mathematical, and data manipulation (fetches, stores, and branches) operations on data elements. In general only a small set of machine language operators are necessary to compute all higher-level programs on any particular computer.

Programs are typically programed in higher-level languages because each higher-level code instruction is equivalent to a whole sequence of machine language operations. This makes coding much more efficient and easier to follow since higher-level variables and subroutines can be assigned meaningful names and programmed in terms of those names.

The various declared data types are all actually stored as binary data bytes with their declared names and types referencing the type of data they represent. Program code is also stored as sequences of binary bytes representing machine language operations with its storage locations providing context that it's code. Storage locations are accessed by pointers to addresses that can contain either data or code routines.

Thus all the bytes of a program are loaded into the fixed memory locations of the computer with pointers and contexts indicating whether the bytes at any particular location are to be interpreted as code or one of the declared data types.

Programs are executed by a processor(s) executing individual machine level operations in a stepwise manner. Conditional branches or calls and returns to and from other code segments are based on push and pop stacks of stored memory locations.

It is almost certainly true that the elemental program that computes the observable universe consists only of the analogue of machine level code. However it seems likely that the emergent programs of the universe can be best explained in terms of higher-level languages programmed not by any programmer but by evolution itself that make calls on the elemental program to compute the processes they encode. DNA is perhaps the clearest example of a higher-level language code programmed by the evolution of the elemental program. But the actions of living organisms also function in terms of calls on a whole hierarchy of lower level routines consisting of multiple elemental operations as for example running involves the coordination of vast numbers of individual cellular operations and muscle coordinations initiated when an animal just decides to run.

COMPUTATIONAL SPACE

Computational space is the logico-mathematical space defined by the computations of the observable universe in the same sense as computer programs define computational spaces. This is the data space in which all the data of the observable universe exists and is computed. It's the data space in which the current present moment entanglement network of all particles, particle components, and their relationships exist. It contains the data of the entire observable universe and is the actual fundamental level of the observable universe.

All the emergent dimensionality of the universe is computed with respect to this computational space and thus actual world lines and rotations are with respect to its parameters so that all actual motions and orientations are relative to it. This computational space exists in the current present moment where the processor of happening continually recomputes all the data that makes up the observable universe. Thus computational space provides an absolute space and time reference background with respect to which all the relativistic dimensionality of the observable universe is relative. The overall distribution of mass-energy structure of the galaxies will be roughly aligned with this array since all mass-energy structures are computed within it.

Computational space can be simply modeled as a 3-dimensional data array whose contents are updated at every P-time tick by interactions of their individual cells though it's not certain this is the actual model used. This array has only numeric rather than physical dimensions in the same sense as an array defined in a computer program does, but it's the background within which the observable dimensionality of the universe is computed.

An important point is this array can be Cartesian (flat) because the curved spacetime of general relativity can be easily modeled with an equivalent flat space where every point has a velocity density. This greatly simplifies the encoding and computations of general relativity in our computational space.

Within this computational space the presence of particles is encoded by valid sets of the particle component data that forms the particle. The particle component values are data in the cell of the array where the particle is located.

Every point of this array has an intrinsic c valued combined spacetime velocity. The intrinsic nature of this computational space is combined spacetime velocity with a total value of c at every point. Space is equivalent to a velocity field, and it's the presence of the c valued velocity at each of its points that manifests as computational space.

Each point in the computational space array has an intrinsic c spacetime velocity that initially is completely through time. We can visualize this by an imaginary velocity meter in which the vertical axis is velocity in time and the horizontal axis velocity in space. Thus initially in empty flat zero-point energy space the velocity meter of every point will point straight up indicating all the c spacetime velocity is only through clock time at that point.

The presence of a charge, for example a mass, near a point tilts the velocity meter arrow towards the horizontal axis proportional to the velocity density field produced by the mass at that point. Because masses and other charges are forms of energy they are equivalent to fields of spatial velocity density by the METc Principle and tilt the STc Principle meter of adjacent points from velocity all in time towards more velocity in space. For the vector totals of space and time velocity at the point to remain equal to c, the velocity of time must decrease. This is indicated by the tilt of the meter.

A mass charge is not just a point but also a spherical field of spatial velocity density that falls off by the square of the distance from the mass. Thus a gravitational field is equivalent to a field of spatial velocity that increases the spatial velocity density of space. This intrinsic velocity density is what slows the velocity of time in gravitational fields in accordance with general relativity.

And because this velocity density field is modeled as fine vibrational waves or pulses in the computational space the actual distance across points in the field is greater because the waves of velocity density must be traversed. This produces a model equivalent to the curved space of general relativity where the greater distance is modeled as traversing the longer curves of space near gravitating masses.

This Universal Reality velocity density model is equivalent to the curved space model of general relativity but much easier to visualize, understand, and compute because it can be represented in terms of a flat space Cartesian array.

This computational space is the underlying stuff of particles because it's a field of spacetime velocity. Charged particles are local excitations of the field of computational space associated with valid particle component sets. The presence of a particle charge is modeled as a tilt in the intrinsic c valued spacetime velocity of points in computational space from all in time to some in space.

Thus dimensional spacetime is a universal field of velocity density with vectors tilted towards spatial velocity when mass-energy is present and mass-energy velocity density fields produce tilts in the spacetime velocity meters of the background computational space.

The correct combined relativistic effects of gravitation and linear motion of particles or objects traveling through computational space can now be computed simply by adding the intrinsic spatial velocity density of the points traversed to their own linear spatial velocity.

And because a gravitational field is a velocity density field the field will contain a spatial velocity gradient that produces a resulting intrinsic spatial velocity vector towards the gravitating mass at every point in the field. This velocity vector is the source of gravitational attraction because it indicates the natural direction and strength of inertial motion in a velocity density field.

If computational space is an array it will have a minimum resolution and background spacetime will be granular at some minimal level far below the scale of particles.

The computational space array can also be hyperspherical simply by curving it over and connecting opposite edges. This eliminates any edges or infinities in dimensional spacetime and neatly models the hyperspherical geometry of our universe as explained in the chapter on Cosmology.

This flat array model appears to be a promising approach to representing the computational space in which the observable universe is actually computed though it needs to be simulated on computers to explore its limits. We have already made some progress towards this end.

LOGICAL OPERATORS

The first category of virtual data in the quantum vacuum is the set of fundamental logical operators happening uses to compute all the programs of the universe. These logical operators are by analogy the 'machine language' of the universe. They encode the most fundamental logico-mathematical, and data manipulation operations upon which all the computations of the universe are based. They are largely equivalent to the operators of Boolean algebra, and the other basic logical, numeric and functional operators of silicon computers.

These fundamental operators most likely include all the essential rules of logic and computation of Turing computers since the reason silicon computer programs have such universal application is they are directly modeled on the actual computational logic of the universe. We can expect equivalent concepts to is, implies, and, or, negation, equals, greater than, less than, stores, fetches, the basic arithmetic operations, etc. to be part of the virtual data of the complete fine-tuning and the computational basis of the elemental program that computes the particle interactions that make up the observable universe.

Sequences of just these few operations are sufficient to generate all possible computations that can occur in the universe. One can get a good general idea of what these might be by examining the basic operator set of silicon computers though there are some important differences as explained in the next section.

These operators necessarily encode an internally consistent system of logic. This means that is they can never generate a logical contradiction. This is critically important to the existence of the universe. If the computations of the universe were not logically self-consistent then contradictions would arise and a computational universe would literally tear itself apart at such inconsistencies and would cease to exist.

The set of fundamental operators must also be logically complete in the sense that they must always produce a result. If they didn't then a computational universe would halt at the incompleteness and also couldn't exist.

Thus the set of fundamental logical operators that is used to compute the evolution of the entire universe must be logically consistent and logically complete. It is likely there is only a single set of viable fundamental operators that satisfies this necessary criterion. This greatly limits the number of possible complete fine-tunings that produce viable universes.

DATA TYPES

The virtual data of the quantum vacuum must also contain some analogue to the basic data types of computer programs. However there will be important differences between the data types of human computational mathematics and those actually used to compute the universe.

For example it's unlikely the string data type would be part of the complete fine-tuning since it's primarily used to store text in computer programs. Another obviously unnecessary example would be the currency data type.

It's not entirely clear to what extent if any abstract numbers divorced from actual instances of things are necessary to compute the observable universe. It could be that abstract numbers may be a human invention extracted from multiple instances of actual things and not necessary to compute the universe. It is possible to perform most arithmetic operations on the basis of comparisons of actual instances of things without using numbers.

For example it appears that the conservation of particle components could be computed on the basis of comparing instances of actual entities to templates rather than counting them, assigning numbers, and then performing arithmetic on those numbers, and applying the results back to the actual entities, a less parsimonious process.

There is also the question of how probability distributions and random choices among them might be made in the absence of abstract numbers, so whether any abstract numbers separate from instances of things are necessary is still uncertain. It may be that the probabilistic relationships among particles might require a specific data type to encode them, or more likely they could be inherent to the operation of the processor rather than stored as data as we suggest in the chapter on Computing Quantum Reality.

For example it's possible to simply assume a degree of randomness to the space versus time allocations of the fixed number of processor cycles used to compute elemental processes as we do to explain quantum indeterminacy below. In this case probability distributions need not be stored as data.

In any case if abstract numbers are used in the computations of reality they are certainly not identical to those used in human mathematics. The number types of human mathematics are clearly generalized simplifications of instances of things encoded in the complete fine-tuning.

For example in neither nature nor a computational reality can there be infinities or infinitesimals. Infinity is not properly a number but a continual unending *process* of adding one forever. This is obviously nothing that actually occurs in reality. Thus infinite or infinitesimal numbers will not be part of the complete fine-tuning data types. If floating-point numbers exist they would have a limited resolution in which case properly scaled integers might be sufficient. Thus none of the irrational numbers used in science to describe phenomena such as the value of π need exist in nature. Such numbers seem be part of the emergent laws of nature useful to *describe* reality at the aggregate scale but are not necessary to *compute* it since all the actual computations of reality occur at the elemental level and have a minimum resolution.

Possibly not even zeros need to exist though it's not entirely clear whether they could be sufficiently coded as absences of things. And likely no reason for imaginary numbers even though they are used to encode wavefunctions since Universal Reality shows how wavefunctions are simply emergent *descriptions* of quantum phenomena and don't actually compute them as explained in the chapter on Computing Quantum Reality.

For anything to be consistently computed it must be exact and therefore digital. Thus the actual data of reality must be stored as bits, numbers or in some digital form perhaps just as the data of the particle components. In any case it must include the data of all particle components, particles and their interactions necessary to explain the observable universe, and must also apply to even the virtual data of the quantum vacuum.

Based on the computer analogy, there is possibly only a single fundamental binary data type consisting of contextually meaningful sets of on and off binary bits. Both the data and code of computer programs are stored as binary bits and just interpreted differently by the processor on the basis of their context. Whether the fundamental subroutines of reality also use binary code is not clear but certainly possible. If so what the fundamental data bits actually consist of with respect to the quantum vacuum is a fundamental question.

It seems that numbers may be necessary to store dimensional values. However integers may be sufficient if their units reflect a minimum scale of granularity. Both positive and negative integers including zero may be necessary to encode dimensional values though this is unclear since all the data of the universe exists simultaneously within the quantum vacuum and thus its processor. Thus array indices may be unnecessary.

Clearly the fundamental data types of the computational universe is a area that needs further work to see what works best in simulations on silicon computers.

ELEMENTAL DATA OF REALITY

Modern physics interprets the *elementary particles* that make up the matter and energy of the universe as the basic components of reality. However it makes a lot more sense to take what science calls the particle properties or quantum states as the true elemental components of reality that actually make up the elementary particles. Thus Universal Reality refers to them as *particle components*. The particle components include particle number, mass and the other force charges, spin, and space and time parity.

The reason that particle components rather than particles should be considered elemental is that they are conserved though all particle interactions while the elementary particles they compose can be broken apart into their particle components and those reassembled into new particles. This occurs frequently when particles interact. Whenever it happens, the total additive amounts of all the particle components remains the same, they are just reassembled into new particles.

Thus it's clear that the particle components are actually the little components of reality necessary to make something real in this universe. They are not so much *properties* or quantum *states* of elementary particles but the actual *components* that make them up and thus make up all the mass-energy structures in the observable universe.

The view that the elementary *particles* are the basic building blocks of reality is an artifact of the outmoded belief in a material universe that remains even though the modern view of particles is far from material in the usual sense. Even standard quantum theory implies that the wavefunctions used to represent elementary particles can't be physical entities since they must be represented by probability waves in a non-physical imaginary space.

Thus Universal Reality considers the particle components the actual components that make up all particles in the universe rather than mere qualities or descriptive attributes. These particle components are the true elemental data structures of the observable universe. Certain valid sets of particle components make up all elementary particles, and thus all the mass-energy structures of the observable universe.

Though this is a completely new interpretation it's a natural extension of the progress of science that led first to the reduction of all material objects to their chemical constituents; then the reduction of chemicals to elements; and finally elements to atoms and elementary particles.

We now take this to its logical conclusion and propose that all elementary particles are composed of an even more fundamental set of particle components that are conserved through all the computations of the universe and that these are actually the fundamental data types of the universe. These particle components are the basic indivisible data types of the universe because it's only they that remain unchanged through all particle interactions. This is precisely the same criterion previously used to reduce all forms of matter to atoms and then elementary particles.

Though the particle components are very nearly equivalent to the particle components, there are a few rare cases in which a few of the currently defined particle components are not conserved individually but only in combination. For example color charges can be switched in weak force interactions (Wikipedia, Quantum chromodynamics) and there are rare cases in which electric charge and spacetime parity are only conserved in combination (Wikipedia, CPT symmetry).

These rare exceptions require an explanation. The exact particle component set needs to be defined as whatever fundamental components are necessary to make up all actual particles, and be conserved through all particle interactions. These will then be the elemental data structures of the universe out of which everything is made.

But even taking these rare cases of conservation violation into consideration particle components are much more strongly conserved than elementary particles and should still be considered the elemental units of reality. The few exceptions are most likely hinting there is an even more elemental unit of data even particle components are made up of, perhaps just binary data bits, that is conserved in all cases without exception.

Or it may be just a matter of properly identifying a set of particle components that is always conserved. In either scenario it will be the elemental set of data units that is always conserved and necessary and sufficient to uniquely identify individual particle types that are the true elemental data units of reality that make up the entire observable universe. So it's always possible there is an even finer binary digital structure that makes up the particle components but for now we assume properly defined particle components are the fundamental data of the universe.

Thus the elemental program of the quantum vacuum that computes the observable universe, is actually computing particle component relationships when it computes elementary particle events.

In Universal Reality the particle components are the fundamental data structures possible in 4-dimensional spacetime, and elementary particles are composed of valid combinations of particle components. A small set of unique possible combinations of particle components forms all known particles. These sets are the basic templates of reality. The particle components are the little bits of the different components of reality necessary to make something real and actual in the form of particulate matter and energy in our universe.

The particle components are specific actualizations of the quantum vacuum that combine according to complete fine-tuning templates in valid sets to form elementary particles. Thus the particle components are the possible forms that can precipitate or crystallize out of the quantum vacuum to form particles under proper conditions.

The observable universe consists of the data of all particle components and their relationships. Their actualized data appears to be all that is necessary for the elementary program to compute the evolution of the entire observable universe.

Why the particle components that exist are the ones that exist, and why the valid particles they form are the ones they form, are both fundamental questions, but it's apparent that this small set and their allowed combinations are required to build a viable universe such as ours, and it's quite possible the set that exists is the only possible set that could exist.

Presumably the particle components that exist are the only possible forms the quantum vacuum is able to produce due to the internal structure of its virtual nature. So the particle components that exist and the particles they form provide a window into the hidden internal structure of the complete fine-tuning.

Contemporary physics interprets the particle components as some of the conserved *quantum numbers* of elementary particles. However the actual necessary and sufficient intrinsic particle components are often lumped together with other quantum numbers, some of which are derivative (hypercharge) or some like flavor perhaps superfluous (Wikipedia, Hypercharge) (Wikipedia, Flavor (particle physics)).

These other quantum numbers are useful in predicting and interpreting particle interactions but aren't necessary to uniquely specify particles. The intrinsic particle components that actually compose the elementary particles will be those sufficient to uniquely and fully determine the identity of particles.

PARTICLE COMPONENTS

It is useful to distinguish between dimensional (spacetime related) and internal particle components. Particle components related to dimensionality are mass-energy and the other force charges, spin (related to rotational symmetry), and space and time parity (handedness in space and time). Internal particle components include lepton and quark numbers.

The particle components include the charges of the four fundamental forces: mass, electric charge, weak isospin charge, and the color charge of the strong force. However mass and the other force charges are all forms of energy which produce the various vibrations and velocity density fields of their forces in spacetime (Wikipedia, Einstein field equations).

The charge particle components of the four forces including mass can be modeled as vibrations of specific forms whose relative motion generates velocity density fields in the surrounding space. These velocity density fields are equivalent to the spacetime dilation or curvature of general relativity as discussed below in the chapter on Computing Relativity. The specific form of the velocity density field corresponds to the type of charge. These spacetime dilation or velocity density fields are the charge *fields* of the four forces. Thus the charge particle components are not points but velocity density fields in the computational background.

Spacetime is the total field of all velocity density distribution. It's produced by the presence of the virtual particles of the zero-point energy and the velocity density fields of all actualized charged particles. The velocity density field of spacetime is a dimensional projection of the numeric dimensionality of the entanglement network. The resulting spacetime consists of specific forms of velocity density fields in the otherwise flat spacetime created by the presence of zero-point energy.

The particle components also include spin, which is intrinsic angular momentum relative to the entanglement network of the universe. Spin is the source of the magnetic force and the poles of the axis of spin are effectively the North and South poles of a little magnet. When the spins of the particles in a material are oriented in the same direction the material becomes a magnet. The intrinsic spin orientation of the computational background is also the source reference for absolute rotation that solves the problem of Newton's bucket as explained below.

Space and time parity are also particle components. They define the particle's handedness relative to the dimensionality of the entanglement network in which the particles are computed. Handedness is reversed in antiparticles, along with the signs of charges. Handedness may be necessary so the spacetime of the universe can turn itself inside out through big bounces though this is speculative. If so the imbalance of particles and antiparticles may be a relict of the last big bounce.

Identity, which physics currently calls number as in lepton and quark number is another particle component. This identity particle component identifies the type of particle the particle component set manifests. Only matter-forming particles have non-zero identity particle components; the force carrying bosons have null identity values. Leptons have lepton number +1, and antileptons have lepton number -1; quarks have quark number +1 and antiquarks have quark number -1.

There is also flavor, which is a quantum number that identifies the 6 leptons and 6 quarks of their three generations of two types individually. However flavor is not conserved under the weak force, which allows leptons and quarks to be converted among generations, so in this sense it probably isn't an intrinsic particle component.

It's also necessary to extend the rest mass particle component to the total mass-energy of a particle since total mass-energy, rather than just mass, is conserved in all particle interactions. The particle components of the other three force charges also carry energy, but the amounts of those charges are independently conserved. In any case total mass-energy is conserved through all interactions, which means that even if forms of mass and energy are converted the total amount of relative velocity is conserved.

Momentum is also conserved through all particle interactions and in this sense is similar to a particle component since the momentums of individual particles can change through events just as total mass-energies can. How velocities (momentums) are used to express the conservation of total mass-energy is explained in the chapter on Computing Relativity.

It should also be noted that linear kinetic energy and momentum are frame dependent and conserved only in the same frame. When observer frames are switched energy and momentum are not conserved because the relative motion of the new frame constitutes energy itself and changes the total energy of the system by adding the energy of its own spatial velocity to that of the process under consideration. When this is taken into account total mass-energy is even conserved across all frame changes.

If we also consider the spacetime 4-position relative to the aggregate entanglement network, the background array of computational space, as a particle component then we have a unique identification of all separate particles in the universe and a way to tell different particles of the same type apart. Otherwise all particles of the same type are identical and cannot be distinguished. This may be necessary to identify and track individual particles in a computational universe though it's not clear there is any need to track individual particles from P-time tick to tick. All the data necessary to compute them presumably exists with the particles in the current present moment P-time tick.

Particle components presumably exist as free virtual data elements in the reservoir of the quantum vacuum in an unassociated state. Enough energy, enough relative motion, can knock them in valid sets out of the quantum vacuum to form actualized elementary particles. These particles actualize out of the virtual flux of particle components and become part of the observable universe carrying the relative motion that created them.

Particle component values may be positive, negative, or zero, except for mass-energy, which can only be positive. And all particle components other than mass can have only a few small proportional values.

The total additive sums but not numbers of particle components are conserved. Thus it's their additive sums rather than their numbers that are elemental so in this sense particle components behave more like numbers than physical entities. They seem to be arithmetic data rather than physical entities, which supports our theory that everything is data at the elemental level.

For example very high-energy particle collisions generally create large showers of many new particles made up of even greater numbers of new particle components. Such high-energy events can create many new pairs of positive and negative particle components out of the quantum vacuum so long as their additive totals are conserved through the event.

This creation of many new particles rather than one or a few very high-energy particles allows the programs that compute them to limit the energies of individual particles as necessary so as not to challenge the c relative motion speed limit of individual particles. It becomes more efficient to distribute the large amounts of energy involved among more new particles instead. Nature tends to distribute high energies rather than concentrate them. This appears to be due to the principle that entropy tends to distribute energy (relative motion) uniformly throughout the observable universe as opposed to concentrating it.

Note also that the basic particle component values (other than mass) should probably be redefined as integral whole units. Physics currently notates some particle component values as fractional but this is mainly a historical convention due to the fact that integer combinations of these fractional units were originally thought to be their base units. An example is the fractional electromagnetic charges of quarks, which were only discovered after whole units of electromagnetic charge had been assigned to the protons and neutrons that are actually three-quark combinations (Wikipedia, Elementary charge).

FUNDAMENTAL CONSTANTS

Another aspect of the complete fine-tuning is the nature of the fundamental constants of nature, the constants not known to be consequences of anything more fundamental. Why does the specific set of fundamental constants exist and why do they have the particular values they do? Are they completely independent or are they interrelated in any way, and can they be derived from anything even more fundamental? And must these constants be exactly as they are to produce a viable universe or could there have been a different set with different values?

It's just these apparently independent fundamental constants that physicists generally refer to as the *fine-tuning* of the universe, but as we have just seen there is much more virtual data necessary to compute a functional universe. This is why Universal Reality prefers the term *complete fine-tuning* to include all aspects of the quantum vacuum irreducibly necessary to the existence of the observable universe as it is, and this obviously includes much more than the values of the fundamental constants.

In fact there is not even a scientific consensus as to the agreed list of what the fundamental constants are. They are often taken as the 26 or so free parameters of the Lagrangian equation of the standard model but this is clearly not the complete list as the standard model is known to be incomplete, and the number of its free parameters may be reducible as well (Wikipedia, Fine-tuned Universe).

In Universal Reality the unified data structure of mass-energy and spacetime is the complete observable universe, and we must clearly include the fundamental constants necessary to explain all aspects of this unity. These seem to fall into at least two general groups, dimensional and non-dimensional constants.

It's also important to note that the actual values of the constants used to compute the universe will most certainly not be those science uses to describe them. This depends on the system of units used to compute the observable universe and there's no reason to believe reality uses the metric system. The speed of light has a different value in meters per second than it does in miles per second. So we must think in terms of the intrinsic strength of dimensional constants such as the gravitational force as opposed to the numeric value of G in any particular set of units and look towards the possibility of a more natural fundamental system of units.

At least some of the fundamental constants are expressible in terms of dimensionless ratios that might be those actually used by reality. The values of dimensionless constants are independent of the choice of units and in this sense more fundamentally meaningful. One can also choose combinations of constants to arrive at dimensionless constants, or play with the units as is done with the Planck units.

Planck was able to derive a set of units of measurements by setting the values of 5 basic constants of nature: the speed of light, the gravitational constant, the reduced Planck constant, the Coulomb constant, and the Boltzmann constant all to 1 (Wikipedia, Planck units).

Using these values a basic Planck unit for length, mass, time, charge, and temperature can be derived. Because the Planck length and time are very small physicists have developed the unfortunate habit of assuming the so called 'Planck scale' is the minimum scale of granularity of the universe but there is no reason to believe this is true as the Planck mass is larger than the mass of elementary particles.

The importance of the Planck constants is they establish a set of natural units seemingly intrinsic to the universe, which simplify many of the equations of nature by eliminating constants from them. So it's likely that computational reality does use some sort of similar set though likely not the Planck set.

In any case the fundamental constants are integral to computing the structure of the entanglement network, which is the data structure of the history of the observable universe. So whatever set of fundamental units works best to compute the universe will likely be used by reality.

The ultimate goal is to reduce the number of fundamental constants of nature as much as possible. Universal Reality takes steps in this direction by redefining gravitational force as fields of intrinsic spatial velocity. Thus to determine the relativistic effect on a particle moving through a gravitational field we can simply sum the total amount of relative motion a particle experiences and take gravitation and the gravitational constant out of the calculation.

Likewise Universal Reality's suggestion of a possible relationship of c to the zero-point energy and the total mass of the universe promises to take us further towards this goal.

THE PLANCK CONSTANT

The Planck constant h is the quantum (minimal unit) of action, which is energy over time, or equivalently momentum over distance. Almost every quantum equation contains the Planck constant to express the minimal values energy or other variables may take (Wikipedia, Planck constant).

In Universal Reality energy is simply spatial velocity, and space itself is the presence of intrinsic c valued velocity so that the Planck constant defines a minimal unit of spacetime, which expresses the granularity of the observable universe and of dimensional spacetime itself. This is related to how the processor computes quantum events, as explained in the chapter on Computing Quantum Reality.

ZERO-POINT ENERGY

The value of c, misleadingly called the speed of light, is the fundamental fixed rate of the velocity of everything in the universe through spacetime. In any particular relativistic situation this velocity is distributed between velocity in space and velocity in time so that their vector sum always equals c.

The particular value of c can be thought of as a function of the fundamental density or resistance to relative velocity of spacetime, its intrinsic resistance to motion through it. This resistance to motion is what restricts all velocity through spacetime to a finite non-zero value.

Flat space is likely a direct manifestation of the presence and strength of the quantum vacuum zero-point energy. Since flat space is the basis in which all spacetime velocity is through time, it's reasonable to suspect the value of c may be related to the value of the zero-point energy if that determines the intrinsic density or resistance to motion through flat spacetime. Thus the zero-point energy may be a measure of the maximum allowable velocity in our universe just as c is.

The zero-point energy can be thought of as the residual observable energy of the quantum vacuum out of which all the mass-energy of the big bang actualized thus the current value of the zero-point energy may also be related to the total mass-energy content of the universe. Thus the zero-point energy value may also be a measure of the curvature of the universe, which is a function of its total mass-energy density. This also suggests a possible relationship of the value of c to the total mass-energy of the universe.

The zero-point energy is also possibly related to the cosmological constant that determines the rate of Hubble expansion of the universe since that is most likely due to repulsive gravitation due to the energy density of empty intergalactic space (Wikipedia, Cosmological constant).

Thus it seems likely there may well be a relationship among the values of the zero-point energy, the speed of light, the cosmological constant, and the total mass-energy content of the universe. Thus all these four fundamental constants could be expressible in terms of a single even more fundamental constant, most likely the total mass-energy content of the universe.

The total mass-energy content was presumably produced by the big bang, likely setting the zero-point energy value as its residual energy. This would in turn set the value of the speed of light in spacetime and the allocation of processor cycles computing it. And in turn the value of the cosmological constant would be a function of these processes.

And since the four fundamental forces can be modeled as curvatures or velocity densities in spacetime, there could also be a relationship to the strengths of the four forces and the particle masses as well as these are also integrated aspects of a single spacetime mass-energy equivalence.

One of the outstanding problems of the complete fine-tuning is the disagreement of over 100 orders of magnitude between measured values of the zero-point energy, which are consistent with general relativity, and those predicted by quantum field theory under various assumptions in accord with the standard model. This discrepancy, called the vacuum catastrophe, clearly indicates a major problem with the standard model whose resolution may well lead to additional insights into the values and relationships among the fundamental constants (Wikipedia, Cosmological constant problem).

It's also important to note that the zero-point energy is the fundamental source of the Heisenberg uncertainty principle (Wikipedia, Zero-point energy). The fact that complementary variables of a system, for example its position and momentum, cannot be simultaneously specified with unlimited precision is due to the fact that all quantum systems exist in the non-zero energy of the quantum vacuum. For the complementary variables of a system to be specified to unlimited precision requires the system must have a ground state energy of zero, but since all systems exist in the quantum vacuum this is never true.

Thus there should be a relationship between the value of the zero-point energy and the Uncertainty Principle's Planck constant minimum precision of the complementary variables of all physical systems that is computed into the entanglement network as it's generated.

PARTICLE MASSES

The distribution of particle masses is one of the great mysteries of the complete fine-tuning. It is precisely because particle masses don't come in exact multiples that particle interactions create spacetime. For particle transformation events to occur some of the vibrational velocities of mass must be converted into linear relative and/or wave frequency velocities, which require a dimensional spacetime to occur within, and so spacetime has to be computationally created to conserve total mass-energy in events. Spacetime has to exist and must be created to allow particle events to occur due to the non-proportionality of particle masses. Thus the non-proportional values of particle masses are the ultimate source of spacetime and the observable universe in which we exist.

The fact that there are precisely three generations of two types of both leptons and quarks, the two categories of matter forming particles, has to be more than a mere coincidence. It has to reflect some deep fundamental symmetry. It's also significant that leptons appear to be free fundamental particles on their own but quarks can only exist in combination in protons, neutrons and mesons.

The fact that the only difference between particles of different generations of all types is in their masses makes this correspondence even more significant.

Amazingly the generational differences in mass of both leptons and quarks are in accord to great accuracy with a simple but little known equation called the Koide formula (Wikipedia, Koide formula) and its extensions (Goffinet, 2008), which virtually clinches their significance. (The masses of the neutrinos and some quarks are not yet known with sufficient accuracy to confirm their compliance but they are within the correct ranges.)

Also the basic three generations of two types template immediately suggests a possible relationship with the three dimensions of space and one of time. It suggests there might be one generation of particles for each dimension of space, and possibly two types corresponding to the two aspects of time, clock time and P-time. However this is currently speculative with little if any supporting evidence.

However the Koide relationship among generational masses does recall the Pythagorean formula of the STc Principle relation of the squares and square roots of space and time variables and adds circumstantial evidence to this assumption.

The Koide relationship states that the ratio of the sum of the masses of the 3 generations divided by the square of the sum of the square roots of those masses is equal to 2/3. This certainly suggests a sort of reverse or inside out connection to the structure of spacetime as expressed in the STc Pythagorean equation. The STc equation involves the square root of the sum of the squares of *space and time* variables, while the Koide relationship involves the square of the sum of the square roots of *mass* variables.

Recalling that mass-energy is equivalent to spatial velocity, and mass-energy and spacetime are two aspects of the same fundamental computational structure as Universal Reality proposes then these two equations may be somehow describing aspects of an underlying symmetry from the opposite perspectives of mass-energy and spacetime.

It is also worth noting that the square root of masses prominent in the Koide formula is how the gravitational effects of mass fall off with distance in 3-dimensional space. Due to the simple geometry of 3-space, the strength of gravitational fields fall off by the square of the distance or equivalently by the square root of the mass which gives the effective gravitational mass at any distance and is thus a basic statement about how mass fields manifests in 3-space.

This could well be meaningful especially in determining how mass affects spacetime during the dimensionalization of distance. The square root of mass simply tells the entanglement network computations how its effects are to fall off with distance as distance is computed. It's telling the conservation subroutine what form to use as it constructs the velocity density field around a mass.

Thus it seems the mass particle component comes only in values reflecting the way its field strength falls off in 3-space and this restricts the values it can take for the 3 generations of massive particles allowed in a 3-space universe in some manner.

While this idea is certainly speculative it may shed some light as to why there are precisely three generations of all types of massive particles whose masses are related by the manner the effects of mass fall off in the three spatial dimensions that characterize our universe. This is certainly possible when we recall that gravitation is a field of velocity density surrounding massive particles.

The Koide relationship doesn't uniquely determine what the masses of particles within a generation are, only what their relationship must be. The relationship holds for both leptons and bosons but their masses differ. So there is obviously some other as yet undiscovered rule necessary to specify exactly why the four leptons and bosons have different masses but still obey the Koide relationship. We need something else that determines at least the base mass in each generational series and the relationships among the four series.

This is likely related to the fundamental question of why only mass doesn't come in a small set of proportional values like the other force charges, and why there are no negative masses.

For example the existence of the universe is extremely sensitive to the ratio of electron to proton mass. If it were only slightly different atoms, molecules and all the structures of the universe could not exist. They would either fly apart or collapse in on themselves and never form in the first place.

The explanation of the relative masses of the four series of matter forming particles is not clear, but is undoubtedly related to the fundamental constants of spacetime given that mass is a form of relative motion that creates a spacetime dilation or velocity density field. If mass and spacetime are both aspects of the same single entity then there must be a deeper connection between the fundamental constants of both.

It should be noted there is another layer to the 4x3 matrix of lepton and quark generations that contains their antiparticles. The antiparticles in this layer have the same masses but their charges and parities are reversed. As a result they are facing backwards in time and have an opposite handedness in space, and the rotational directions of the helical spacetime dilations or velocity densities of their electromagnetic fields are reversed as explained in the chapter on Relativity And Electromagnetism.

FOUR FORCES

There are four fundamental forces in the universe each with a class of bosons thought to carry the force and act upon its charges. Universal Reality also models these forces as four distinct types of spacetime vibrations and their associated spacetime dilation, or velocity density fields.

Why there are four and only four forces in our universe is possibly related to the manner in which the quantum vacuum must manifest as a 4-dimensional spacetime, and the distinct types of vibration-dilation structures that are possible in 3-space though this is speculative.

There are several different ways to represent the elementary particles and their particle components in charts. One is to add the four forces to a 4x3 matrix of leptons and quarks across the bottom to obtain a 4 x 4 matrix (Wikipedia, Elementary particles).

From left to right we have the recently discovered Higgs boson thought to give all particles their masses. Next the photon that carries the electromagnetic force, then the weak bosons that carry the weak force, and last the gluons that carry the strong force.

This last row is a little different from what is seen in the usual charts, but they are somewhat arbitrary in this respect as there isn't any natural association with the force carrying cells of the standard charts and the particle classes above them. Thus it seems more natural just to add the four forces in separate cells of the chart rather than to first omit the graviton, and then put the two W and Z weak force bosons in separate cells, but lump all gluons of the strong force together in the last cell as the chart illustrated does. That doesn't make much sense.

Each of the four forces has a different number of charges so the actual number of particles in each position in this bottom row increases from left to right. The gravitational force has only one charge corresponding to the existence of positive mass. The electromagnetic force has the two plus and minus electric charges. The weak force has the three isospin charges of 0, ½ and – ½. And the strong force has 6 color charges (±red, ±green, ±blue).

Particle physics considers the four forces to be mediated by exchanges of particles called bosons. The strong interaction is mediated by exchanges of 8 types of gluons, the weak interaction mediated by exchanges of Z and ±W bosons, the electromagnetic force by exchanges of photons, and gravitation by exchanges of the hypothetical graviton.

However general relativity models gravitation as fields of spacetime dilations, and Universal Reality proposes the forces are actually different forms of vibrational velocity density equivalent to different types of spacetime dilation. This is a good conceptual model but since Universal Reality considers spacetime to emerge from particle component interactions the two models must have an equivalence that arises *in the relative scales and forms* of the entanglement networks surrounding charges of the various forces.

As the entanglement network is generated by particle component interactions it's given a relative scale and dimensional form by the force interaction involved. This can be interpreted as vibrations and spacetime dilations of forms corresponding to the four forces as outlined in the chapters on Computing Relativity and Relativity And Electromagnetism.

Thus our chart takes the four forces as primary, and the numbers of their charges and force carrying bosons vary increasing from left to right just as the masses of the quarks and leptons increase from generation to generation.

Our matrix also associates the recently discovered Higgs boson with the gravitational force, rather than the hypothetical graviton. The graviton is not even part of the standard model and doesn't appear on the usual particle charts but the existence of the Higgs is confirmed and part of the standard model. The Higgs is the particle thought to give all particles their masses in the standard model so it's the closest thing the standard model has to a graviton. And it is imagined as a field that pervades all of space as part of the quantum vacuum so it's clearly related to how the constants of spacetime and mass-energy interrelate.

In Universal Reality spacetime is actually a field of mass-energy including the ubiquitous zero-point energy of the quantum vacuum. And mass-energy is excitations in this field including the ubiquitous zero-point energy excitation of the whole field. In particular all the particle charges are localized excitations in the mass-energy field of spacetime.

This is the equivalence of mass-energy and spacetime in a single computational structure. And this is why the fundamental constants of both mass-energy and spacetime must necessarily be related.

Thus the values of the zero-point energy, c, G (the strength of gravitation), the strength of the other forces, and the masses of all particles due to the Higgs field are almost certainly related since they are all constants of a single mass-energy-spacetime equivalence.

And more generally it is quite likely that the three generations of massive particles correspond in some subtle manner to the three dimensions of space, and perhaps bosons have some symmetry connection to the time dimension. Perhaps the matter forming leptons and quarks are more a manifestation of spatial excitations or symmetries, and the force carrying bosons more of symmetries in time?

In Universal Reality all forms of mass-energy are forms of relative velocity or motion in space. So the values of the constants of the strengths of the four fundamental forces, must be proportional to the amounts of relative motion each carries. For all forms of mass and energy to be convertible they must all be forms of the same underlying thing, relative motion in space. Thus the values of the constants of the force strengths must be proportional to the amounts of relative spatial velocity those forces carry.

FOUR DIMENSIONS

It is fairly clear that there can only be 4 dimensions in a viable universe such as ours. There must be three space dimensions because it takes at least three for meaningful structures to exist, and one and only one dimension of clock time for things to happen, and to happen in a consistent manner.

For example in a two dimensional space beings with internal digestive tracts couldn't exist because having one would split the being into two separate parts. For the same reason most types of compound structures with internal components couldn't exist and no complex life could exist. More fundamentally there couldn't even be any atoms or molecules in a 2-dimensional space since these are all 3-dimensional structures and there are no 2-dimensional equivalents. Basically there is simply not enough room in two dimensions for any meaningful structures to exist or evolve. Entities would always be bumping into each other and interfering with the movement and structures of other entities as there would be no third dimension for anything to go around anything else.

And in a universe with more than three spatial dimensions other types of meaningful structures cannot exist. Knots for example can only exist in 3-dimensional space. Stable planetary orbits can't exist because the strength of gravity would vary much more rapidly with distance and small perturbations would cause planets to quickly fall into their stars or escape their orbits. It's unlikely the basic laws of physics and chemistry produce viable mass-energy structures in more than 3 dimensions.

There must be at least one dimension of clock time for events to be able to occur at all. However two or more dimensions of clock time would allow physical processes to run at different rates or towards different futures in the same location and that would immediately lead to logical contradictions that would tear the fabric of a computational universe apart.

However in addition to the 4-dimensions of spacetime a separate over arching dimension of P-time is necessary in a viable universe to reconcile, contain and logically relate all the relativistic clock times that occur at different locations in a four dimensional universe with a finite speed of light.

Without an STc Principle with some finite speed of light clock time would either not pass at all and nothing would ever happen, or clock time would pass instantaneously and the entire history of the universe would immediately be over before it even began.

Thus only a universe with 3 spatial and 1 clock time dimensions such as ours seems viable, and there must also be a separate present moment P-time, and a single underlying computational space for computational processes to evolve in a consistent relativistic manner, and to exist in the same universal present moment of existence in the same observable universe.

Universal Reality models the charges of the four forces as either different forms of relative vibrational motion and dilation fields in 4-dimensional spacetime or as equivalent particle exchanges of bosons so there's no need for the hypothetical but unverified additional compacted dimensions proposed by String Theory (Wikipedia, String theory).

String Theory proposes vibrations in multiple compacted dimensions to account for different types of particles, but in Universal Reality it's the data of particle components that are the fundamental elements of reality and the force carrying particle component charges are modeled as different forms of intrinsic velocity density fields in standard 4-dimensional spacetime. And these are produced by the conservation of particle components in particle interactions.

Thus it seems reasonably certain that the four familiar spacetime dimensions of the universe along with P-time are precisely what is required to construct a viable universe, and this aspect of the complete fine-tuning of the quantum vacuum can only be exactly as it is. There are likely no viable universes possible with either more or fewer dimensions than our own.

DEEP SYMMETRY

The fact that there are precisely four types of massive particles that make up all matter, four forces that govern the interactions of this matter, and four dimensions in which this matter exists has to be much more than a mere coincidence especially considering the deep equivalence of mass-energy and spacetime as two aspects of the unified computational structure of the quantum vacuum. So if we add the four dimensions of spacetime to our matter and force table we get a beautiful 4 x 4 x 4 unity that summarizes essentially the entire observable universe in a nutshell. Mass-energy and spacetime constitute the entire observable universe, and this 4 x 4 x 4 unity boils it down to its essence.

Clearly we have the glimmer of a much deeper hidden secret here. The fact all these fundamentals come in sets of four certainly hints at a deeper unity. The unity of mass-energy and space and the fact both are composed of four fundamental units must certainly be more than a coincidence.

The force charges of the elementary particles are actually excitations of spacetime itself. The shape of spacetime is the distribution of mass-energy in the universe. Mass-energy is spacetime, and spacetime is mass-energy, and mass, energy, space and time are all forms of the single fundamental element of relative motion or velocity. Everything in the observable universe is part of a single unified data structure computed by quantum events.

Everything takes place in the abstract computational space of the quantum vacuum. It is there that the elemental program computes the interactions of the data of the particle components that make up the universe, and in doing so generates the entanglement network that we observers interpret as our familiar physical spacetime universe.

Since mass-energy and spacetime are both aspects of the single conserved element of velocity it's clear that the fundamental constants of mass-energy and spacetime must be related at a deeper level. We have seen hints of this but there is certainly much more to be revealed.

Thus it seems quite likely there is an undiscovered relationship among the values of the zero-point energy, the speed of light, the cosmological constant, the total mass-energy content of the universe, the nature and charges of the four forces, and the specific particles and particle components that exist. All these fundamental constants are almost certainly expressible in terms of a single elegant set of even more fundamental constants, or perhaps ultimately a single constant or principle that uniquely defines our universe and the computational structure of its mass-energy spacetime equivalence.

It is unclear if, or to what extent, the complete fine-tuning might be altered and still produce a viable consistent universe though probably not much if at all. Many aspects of it, such as the basic rules of logic and the four spacetime dimensions, are clearly necessary exactly as they are and it's quite likely that the complete fine-tuning of our universe is the only one possible.

THE REAL MICROCOSM & MACROCOSM

The templates of all the actual information structures that exist in the observable universe exist in a virtual form in the complete fine-tuning of the quantum vacuum. The computational space of the quantum vacuum is not a dimensional structure but it contains the virtual template for dimensionality and dimensionality is computed within it on the basis of that template. This is true of all aspects of the observable information structure of the universe. They are all implicit in the virtual information structure of the quantum vacuum in which they are computed.

Thus the fact that for something to be real and actual in the universe it must be composed of elemental bits of identity, charge, mass, spin, space and time handedness and so forth indicates that the quantum vacuum itself consists of all these aspects of reality in a virtual form.

The fact that dimensionality has characteristics of position, relative velocity, scale, orientation, etc. indicates that these fundamental elements of reality also exist as virtual templates in the quantum vacuum. They are there in the virtual data of the complete fine-tuning and whenever anything becomes actual it must crystallize into actuality according to these virtual templates.

Thus when particles crystalize out of the computational spacetime background they take on the forms implicit in the computational spacetime from which they crystallize. Particles and dimensional spacetime and their myriad relationships are all crystallizations of the absolute background computational spacetime in which they are computed. They are actualized manifestations of the underlying structure of the complete fine-tuning that exists all around us in a hidden virtual form.

We and all the particulate structures and programs of the observable universe including our dimensional structures, are computed within and exist within the quantum vacuum that supports our existence with its own. We are the direct manifestations of the underlying template forms of the complete fine-tuning that exist within us in virtual form and that buttress our observable structures and processes. Without the actual existence of the templates of the complete fine-tuning within our own being and the being of the observable universe neither we nor the observable universe would exist.

The traditional concept of microcosm and macrocosm is certainly incorrect but this is perhaps the true meaning of the microcosm and macrocosm being reflections of each other. There is only a single kind of thing that exists within the sea of existence of quantum vacuum and that is data given existence and being by its presence.

This data takes an elemental set of highly related virtual template forms in the complete fine-tuning. The quantum vacuum that pervades the entire universe and exists within all things includes the presence of these forms as virtual templates. Thus they exist within all individual things and when things appear they appear according to these forms. Everything that exists is an actualized instance of the virtual template forms that exist within them, and only the continuing presence of these virtual forms in the actualized forms maintains their existence as their data is continually recomputed.

So spacetime itself with its positions, velocities, orientations and clock time rates is a direct manifestation of the continuing hidden presence of the quantum vacuum, which defines what dimensionality is and what characteristics it must have to become dimensional.

The same is true for all particulate mass-energy structures. They are all observable forms constructed around the hidden blueprints that determine what particle components they must have to become actual. Every particle in the universe consists of its particle components built around a hidden instance of the complete fine-tuning template that determines what these actual particle components must be.

Thus in everything that exists there is combined the actual and the virtual, the microcosm and the macrocosm. All the elementals that exist are all manifestations of the universal. The universal exists within every elemental and all the elementals all exist within the universal.

Also all the charges that exist are little elemental crystals of spacetime that crystalize out of the computational background around valid sets of the other particle components necessary to make real and actual particles in our universe.

It is this crystallization process that creates the observable universe of particulate dimensional structures we humans interpret as a material world within spacetime. These particle crystals presumably appear when little units of spatial velocity move relative to their background and seed the process. When that happens the quantum vacuum gives up elemental quanta of the particle components it takes to make an actual particle so there is something to manifest an individual velocity.

We have the minimum units of everything on the one hand and the maximum speed limits etc. of all those things and their universal templates on the other. The myriads of minimum units of the tiny fill in the maximum scale and size of the universe. It's as if there was an original pulling apart or separation of the fabric of existence into the overarching single units of the maximum and the innumerable individual units of the minimum. From this perspective the original differentiation of the formless was not into male and female but into the large and the small, a differentiation of innumerable individual actualized units of the virtual templates implicit in the formless sea of virtual existence.

One can imagine the fundamental forms of existence being pulled out of the quantum vacuum into little actualized examples of those principles. The observable universe consists of innumerable specific instances of the universal principles of the complete fine-tuning and then begins to evolve via the interactions possible to their nature.

Though clearly the specific instances are created from the general principles one wonders if there isn't some possible feedback mechanism in which the probabilistic evolution of the specific instances towards convergent evolutionary ends implicit in the complete fine-tuning somehow feeds back into influencing the general principles of the virtual complete fine-tuning?

Perhaps this could somehow occur when and if the universe turns inside out in a big bounce and entropy and time reverse and the small somehow determine the principles for the new bounce that might possibly be created as explained in the chapter on Cosmology in Universal Reality (Owen, 2016). In this way could the universe perhaps continually evolve its complete fine-tuning through each big bounce towards some ultimate goal of universal self-awareness? There is certainly much to be discovered…

THE ELEMENTAL PROGRAM

We now come to the last category of the complete fine-tuning. This is the elemental program that actually computes everything in the universe. This elementary program consists of subroutines that implement the fundamental laws of nature. It implements the fundamental principles outlined in the previous chapter.

The elemental program that computes everything in the universe consist of a small set of fixed routines that operate on particle component data. These are the elemental laws of nature of the universe that compute all the data interactions of the observable universe using sequences of the basic logical operators to implement the fundamental principles and complete fine-tuning. The elemental program computes the existence of particles, their interactions in particle events, and their relationships in terms of their particle components. These computations are performed on the basis of the principle of conservation of particle components and compute only events that distribute particle components in allowed sets among emitted particles and exclude any combinations that don't form actual particles.

The basic subroutines compute the STc Principle, the Exclusion Principle, the Particle Component Conservation Principle, and the Principle of Random Choice. For completeness the two fundamental principles of existence and logical consistency give consistent existence to the data of all computational results.

There are also subroutines that compute the bound entanglements of the mass-energy structures of atomic and molecular matter. These are based on balances of the fundamental forces and their effects on individual particle interactions in bound states.

There are also routines that compute particle charge velocity density fields surrounding the charges. This includes the gravitational fields produced by mass-energy and the electromagnetic fields produced by electric changes. It is not clear whether the forces carried by the strong and weak forces are best modeled by fields or boson exchanges but these are likely equivalent models as charges are modeled as velocity excitations of space.

Particle interaction events can be modeled as all the particle components of multiple particles being transiently collocated at the same point. This triggers the exclusion principle, which disallows particle component combinations that don't form valid particles. As a result the conservation principle is triggered which distributes all particle components among valid new particles that are then ejected with the left over energy conserved in the event mainly as resulting linear velocities.

Thus events are not data entities but computations. So they don't properly have frames of their own and are all computed within the background frame of the quantum vacuum.

This single elemental program is the only program that exists in fixed virtual form in the complete fine-tuning and it can be said to make up all emergent programs and compute the entire universe. All the emergent programs of the observable universe compute by making calls on the elemental program and its subroutines to compute the interactions of their constituent particles. By analogy the elemental program is the fixed firmware of the universe.

All the *emergent level* laws of nature that are the subject of most of science *describe* but don't actually *compute* the structures and behaviors of emergent level processes. All emergent processes are actually computed by organized mass calls on the elemental program to simultaneously compute every one of their particle interactions at each P-time tick.

Thus all the emergent programs running in the universe from atoms and molecules on up are aggregate manifestations of these fundamental logico-mathematical routines. All emergent programs consist only of ordered data structures produced by the elemental programs, which in turn are executed in terms of a small set of logico-mathematical operations by analogy the machine language of the universe.

These fundamental subroutines appear to be rather simple, much simpler than the plethora of emergent laws that describe but don't directly compute aggregate behavior. This small set of relatively simple computational routines seems to be all that is necessary to compute the entire evolution of the universe.

More work needs to be done to nail down the precise details of the complete fine-tuning and the elemental program. Ultimately this is probably best done with computer simulations to see what most effectively models the fundamental computational processes of the observable universe. We have made some progress in doing this on the XOJO programming platform.

HIGHER-LEVEL LANGUAGES

Because they operate by making calls on the elemental program the emergent programs that compute the vast majority of processes in the observable universe can be thought of as examples of higher level programming languages evolved by reality itself.

For example all the processes of chemistry and biology, though all actually computed in terms of elemental logico-mathematical operators do consist of more or less fixed calls on sequences or sets of the elemental operators. Thus each of those calls could be considered a higher-level language operation calling subroutines of machine language operations.

The programs of mind's construction and operation of the simulation are another clear example of an evolved higher-level language. Organisms' simulations of their environment have evolved through the progression of species from the beginnings of life. Basically these all operate in terms of a higher-level language we can call *the logic of things*, the basic rules of how individual things and processes are extracted from the perceptual background and work at the classical level. Organisms function within their environments on the basis of the logic of things, which is effectively, a higher-level language ultimately implemented via calls on the elemental program.

Thus the emergent laws of science that describe but don't actually compute the processes of the observable universe are effectively descriptions of elemental processes in terms of the higher-level language we call science. Though most of the programs that compute the processes described by science are just emergent manifestations of elemental level computations, it's also clear that these elemental processes have evolved higher-level programs in the form of living organisms, which operate intentionally to foster their function and survival. These programs operate in terms of the higher-level programming language of their simulation to direct their actions, which are all implemented by calls to the elemental programs that actually compute reality.

These purposeful programs have evolved because by optimizing their actions in support of their functioning they tend to be selected and perpetuated over those less successful. Thus, based on the mysterious rules of the complete fine-tuning, the elemental program itself naturally evolves higher-level purposeful programs in the form of living organisms.

Perhaps the clearest example of a higher-level language evolved by the universe is DNA. DNA clearly is implementing a higher-level language, ultimately based in calls to the elemental program, which codes for the creation, growth, and functioning of biological beings and consists of coded instructions in that language. Many known emergent processes can also be modeled as higher-level languages and we should certainly look for others as well.

All these emergent programs and the higher-level languages they implement are programmed by the general principle of evolution and reused over and over because they are adaptive results of the elemental program. They can be thought of as separate programs, though actually manifestations of the elemental program in the same sense that emergent patterns automatically arise from the elemental operations of cellular automata (Wikipedia, Cellular automaton).

UNDERSTANDING TIME

THE NATURE OF TIME

The central and most fundamental experience of our existence is our consciousness in a present moment through which time flows.

The present moment is simply the presence of reality, the presence of the universe. For reality to exist it must be present and its presence manifests as the present moment in which we experience our existence. Thus our experience of existing within a present moment is our direct experience of the presence of the universe within us and ourselves as a part of the universe.

This means the entire universe exists within the present moment it creates by its presence. Thus there is a universal present moment in which everything exists and there is nothing that exists outside this universal present moment because there simply is no outside. The universe and the present moment it manifests by its presence are both collocated and ubiquitous. They are all that exists. Outside the present moment of the universe there is not even nothingness. The universe in the present moment is all that exists.

Thus the past is a non-existent logical projection inferred backwards from the present. It exists only in memories and its other results in the present moment. And the future doesn't exist because it has not yet been created. Reality exists only in the present moment manifested by the presence of existence.

Note that some physicists still deny the existence of a present moment because they believe it's inconsistent with relativity but this is based on a misunderstanding of relativity as explained below. There is no doubt whatsoever that a present moment exists because it's the most fundamental and persistent of all observations both scientific and personal. The crux of scientific method is to develop theories that explain observations, never to deny them. Denying observations is the antithesis of science and the existence of the present moment is the most fundamental of all observations.

In contrast to the present moment the time that flows through the present moment, what we call clock time, is simply our experience of the continual happening of the universe. Events continually happen in the universe, processes occur and the universe evolves.

For the universe to observably exist it must continually happen. Things must change and processes must continually occur to register as experiences. Our consciousness of the flow of time depends on things changing within a flow of time because only comparison to prior states makes things observable.

Happening is what brings the universe to life and gives it reality in the present moment and sets things and processes into motion within it. Thus happening is an essential element of the existence and reality of the universe and is the source of what we call clock time. Without happening continually occurring you would not be here reading this word. Your experience of the flow of clock time through the present moment is your direct experience of the fundamental process of the universe occurring within your own being.

The flow of clock time is not really so much a separate independent entity that carries individual things along with it but is simply the rate at which processes occur. There is no independent flow of time other than the rate at which processes occur. Thus clock time is not really a thing in itself but is abstracted from the rates at which processes occur.

When processes occur they always happen at fixed rates relative to each other and those rates are what we call the flow of time. Thus the specific rate at which processes occur in any given location is referred to as the local clock time rate at that location.

What this means is that the clocks that measure clock time are simply devices designed to run at standard rates and they measure time by comparing the rates of other processes to their own standard rate based on the principle that local processes always run at fixed rates relative to each other. So clock time is always the rate of some process relative to the rate of some other process. It's intrinsically relative rather than absolute.

For example our personal experience of the flow rate of clock time is relative to the rate of our biological clock, which can vary slightly depending on circumstances. This is why we may experience differences in the rate at which clock time seems to flow at different ages and in different emotional situations.

This also means that without some process occurring there is no flow of time. Thus before the big bang there was presumably nothing happening and thus no flow of clock time and clock time wouldn't have existed (Hawking, 1998).

So we can say that the flow of clock time is simply a term for the rate at which the processes of the universe are being computed at any particular location. If we think of the universe as a computer that continually computes all the processes that are happening within it we could say the processor rate of that computer would determine the clock time rate of the processes it was computing.

Keeping this in mind it's still very useful to speak of the flow of clock time as if it were some actual independent thing even though we are really just talking about the rate at which processes are occurring.

Thus we can say that the present moment is the manifestation of the presence of reality, and the clock time that flows through the present moment is simply the rate at which the processes of reality are happening. These two points explain the fundamental nature of time.

TIME AND RELATIVITY

One of the great discoveries of relativity is that clock time flows at different rates in different circumstances. Specifically in the presence of spatial motion clock time slows. This is referred to as 'time dilation' in relativity and has been extensively verified by observations and must be taken into consideration in space flight and GPS computations (Wikipedia, Global Positioning System).

This means that the rate at which things happen depends on how much relative spatial motion they are experiencing. Things moving faster in space will have their time slowed, and things moving close to the speed of light will have their time slowed to a crawl. However nothing can move faster than light so the speed of light is an absolute limit to all motion in the universe.

The slowing of time with spatial velocity is given by a simple equation called the Lorentz transformation (Wikipedia, Lorentz transformation).

$$\frac{dT}{dt} = \sqrt{1 - \frac{v_x^2}{c^2}}$$

In English this says that the relative time rate of a clock moving with spatial velocity v_x slows proportional to v_x^2 divided by c^2 where c is the speed of light. Thus since most spatial velocities are very small compared to the speed of light, time doesn't slow very much until spatial velocity begins to approach the speed of light. This is why we don't typically see clocks slowing as they move around on Earth. However the slowing is measureable aboard the ISS and spaceflights in general and for any process that has a significant spatial velocity.

THE STc PRINCIPLE

Now there is a largely unrecognized fundamental principle that underlies relativity from which the Lorentz transform is derived [1]. This can be called the STc (space, time, speed of light) Principle and states that the combined space and time velocities of everything in the universe is always equal to the speed of light c.

The equation for the STc Principle is

$$v_x^2 + v_T^2 = c^2$$

which states in English that the square of velocity in space v_x plus the square of the velocity in time v_t is always equal to the speed of light squared.

Note that the combined velocity through space and time is a vector sum rather than the simple addition of their speeds. This is because time is considered a 4th dimension orthogonal to the x, y, and z dimensions of space. Thus by the Pythagorean principle of vector addition it's the square root of the sums of the squares of time and space velocities that's always equal to c [1].

To understand the STc Principle we need to understand what is meant by velocity through time. Relativity expresses velocities through time as *relative* clock rates times the speed of light. Multiplying v_T by the speed of light puts velocity through time in the same units as velocities through space and enables them to be correctly compared as velocities through different dimensions of a single 4-dimensional geometry.

By definition the relative clock rate of an observer's own clock is 1. So multiplying 1 times the speed of light gives c as the velocity through time of an observer and his clock. Thus every observer is always moving through time at the speed of light according to his own clock. The speed of light is actually the velocity in time of everything in the universe including ourselves on its own comoving clock. We, and everything in the universe, are always moving through time at the speed of light on our own clocks. This is what we experience as the passage of clock time.

By definition observers don't move relative to themselves. This is why all the c spacetime velocity of every observer is completely through time on its own clock. Thus c is the speed of time on all observers' own clocks and this is also why light always seems to travel at c for all observers.

Relativity tells us that a clock moving in space relative to us will seem to be running slower than our own clock. Thus two observers moving relative to each other will both see each other's clock running slower than their own. The amount of slowing depends on the relative spatial velocity and is given by the STc Principle equation.

The STc Principle states that the combined vector velocity through space and time of everything in the universe is always equal to the speed of light, c. This is a consequence of relativity that is well known to scientists though usually viewed as a mere curiosity (Greene, 1999, 2005). However it's actually a fundamental principle with profound consequences. It means that every clock in the universe runs slower in time proportional to its velocity through space so that the combined spacetime velocity through time and space of everything in the universe always remains equal to c.

This is one of the fundamental conservation principles of reality. It means that the total velocity of everything in the universe is always c, and this velocity is distributed between velocity in space and velocity in time. Thus if velocity in space increases the velocity in time must decrease so that their vector sum always remains equal to c. Thus c is the fixed velocity of all processes in the universe. Everything in the universe must always move with velocity c no matter how that velocity is distributed between velocity in time and velocity in space. This is a fundamental law of the universe. The STc Principle is the single essential key to really understanding relativity and its effects including how time works.

In a computational universe the STc Principle can be envisioned as the consequence of a fixed number of processor cycles used to compute the happening of every process in the universe. Processor cycles go first into computing the spatial velocity of a process. Then the remaining cycles are used to compute the internal evolution of the process, which manifests as its clock time rate. In this manner the processor cycles that compute each process are distributed among computing velocity in space and velocity in time so that their vector sum is always c. This processor model is explained in detail in the next chapters.

The rate of internal change of state of any process manifests as its associated clock time rate, so the clock time rate slows as a function of the computation of relative spatial motion. This automatically manifests as the STc Principle, which describes how clock time works for all processes. The processor cycles of happening are distributed between computations of relative spatial motion and the internal clock time rates at which processes occur. This is how the STc Principle emerges from the computation of processes in a computational universe.

The STc Principle means the speed of light is actually the intrinsic velocity of spacetime and light just happens to move at that velocity through space because it has no internal processes to be computed and thus no internal velocity through time. If light had a comoving clock its hands would never move. Thus understanding the STc Principle explains why light itself always has to move through space at the speed of light. It's because its own internal clock isn't moving at all in clock time so all its c spacetime velocity is through space.

The 3×10^8 meters/sec value of c in our universe provides enough time for things to happen and enough space for things to happen in. If the speed of time was zero nothing could ever happen, and if it was infinite the whole history of the universe would be over before it began, so a viable universe requires a reasonable finite non-zero value of c to be encoded in the complete fine-tuning of the universe.

The value of c must also be quite large, as it is, relative to typical velocities through space. If it weren't the spacetime dilations of mass-energy would produce gravitation so intense as to crush all possible material structures, and routine spacetime distortions so great as to make ordinary processes unintelligible.

The STc Principle not only underlies most of special relativity but also provides a firm physical basis for the apparent mystery of the arrow of time and confirmation of a privileged present moment by relativity itself, a fact that neither Einstein nor most other physicists seem to have recognized. This will be explained shortly.

The STc Principle applies only to clock time; it doesn't apply to P-time, which has no observable rate since it's an intrinsic aspect of computational reality, namely the processor cycle rate that is fixed throughout the universe.

TIME & GRAVITATION

Relativity tells us that time also slows in gravitational fields which suggests gravitation might be some form of spatial velocity. In fact it's easy to model gravitation as a field of vibrations in space whose intrinsic velocity is equivalent to their gravitational force (Owen, 2016).

This is a simple and elegant theory that allows both linear velocity and gravitation to be seen as two aspects of the same phenomenon. Basically the theory says that all forms of mass and energy are different forms of spatial velocity. This has the added advantage of explaining how all forms of mass and energy can be conserved as science has shown they are. It's easy to understand how various forms of the same thing can be converted from one to the other and in fact the only way anything can be conserved is if it's all forms of the same thing. Thus we discover the underlying reason for the conservation of all forms of mass and energy. They are all forms of spatial velocity.

Thus the conservation of mass and energy is simply the conversion of equivalent amounts of spatial velocity among its various possible forms. The linear velocity of motion, the wave frequency velocity of electromagnetic radiation, and the vibrational velocity of gravitating mass are all interconvertible forms of spatial velocity. This insight makes understanding time and its relativistic effects much easier to understand.

To understand the STc Principle in the context of gravitation we simply model a gravitational field as a field of intrinsic velocity in space. Thus an object seemingly at rest in a gravitational field actually experiences an intrinsic spatial velocity because it's in a velocity density field.

A gravitational field is a field of intrinsic velocity with strength inversely proportional to the square of the distance from the mass that produces it. Thus a gravitational field has an intrinsic velocity gradient extending out from the mass and every point in the field has a resulting velocity vector towards the mass because the intrinsic velocity of the field is greater closer to the mass than away from it and the difference at any point produces a velocity vector pointing towards the mass.

Thus an object at rest in the field experiences a velocity vector towards the mass that its inertial motion tends to follow. This is the source of what we call gravitational attraction. There is no actual attraction in the sense of a pull. It's just a matter of an object in inertial motion traveling through space along its velocity vectors.

We confirm a gravitational field as a velocity field in our direct experience as the velocity we experience falling towards a gravitating mass along the velocity vectors it produces. And the force of gravity we experience standing on the surface of the earth is actually an acceleration against this inertial motion due to the surface of the earth blocking our motion. This is in agreement with Einstein's Equivalence Principle that states that what we feel as a gravitational force is actually an acceleration (Wikipedia, Equivalence principle).

So to understand how the STc Principle works in a gravitational field we simply add the intrinsic velocity density of the field to any linear velocity an object has. The velocity through time of the object then slows due to the combined spatial velocity of linear velocity and the intrinsic gravitational velocity of the field. This explains why both linear velocity and gravitational fields produce time dilation. They are both forms of spatial velocity that slow the velocity of time by the STc Principle.

Thus a clock at rest in a gravitational field will observe a clock at rest in empty space running faster because its own clock rate is slowed by the intrinsic velocity of the field. The clock in empty space will likewise see the clock in the gravitational field running slower for the same reason. Both observers agree on this effect; thus it's an actual relativistic effect and the difference in elapsed time produced is permanent.

The velocity density model explains the relativistic slowing of time by gravitational fields in the same way time is slowed by linear velocity. Gravitational time dilation and the time dilation of linear motion are both due to increased spatial velocity slowing velocity in time in accordance with the STc Principle.

Thus the total spatial velocity of an object is its linear velocity plus the intrinsic velocity of any fields it's in. This gives its total spatial velocity, which combined with its temporal velocity must always equal c.

Thus we get a very simple way of understanding general relativity that works for all observers in all situations. An observer in a gravitational field just needs to add the intrinsic velocity of the field to any linear velocity he may have to correctly calculate his time dilation compared to an observer at rest in empty space.

This velocity density model of gravitation is equivalent to the curved spacetime model of general relativity and predicts exactly the same effects but is superior because it provides an actual mechanism for those effects that is lacking in relativity. Relativity provides no mechanism for how the presence of matter curves spacetime, it just does. However in the velocity density model mass and its gravitational field *actually are* fine spatial vibrations whose spatial velocity works seamlessly with the STc Principle. Gravitational mass and linear velocity now become two aspects of the same thing as the conservation of mass and energy suggests they should have been all along.

And not only that the velocity density model has the additional advantage of modeling space as we actually see it, as a flat uncurved Euclidean space rather than the curved space of general relativity which we never observe. In the velocity density model space is flat and uncurved as we actually see it, it just contains areas of intrinsic velocity density around masses because masses actually are fields of velocity density in space.

THE VELOCITY FABRIC OF SPACETIME

The STc Principle is a universal fundamental principle that tells us the speed of light is actually the fundamental velocity of spacetime itself, and space itself can be considered as nothing more than a field of intrinsic velocity whose value is c at every point. In this view the fabric of space itself consists entirely of intrinsic spacetime velocity.

The amount of that intrinsic velocity that consists of intrinsic *spatial* velocity is the strength of gravitation at any point in space. Correspondingly the amount of total velocity at any point that is not the intrinsic spatial velocity of a gravitational field is the intrinsic velocity of time at that point. This gives us a very neat and elegant picture of the basic unity of mass-energy and spacetime that automatically obeys the rules of general relativity which can be called the METc (mass, energy, time, speed of light) Principle (Owen, 2016).

Thus the speed of light should actually be understood as the intrinsic velocity of spacetime, or more fundamentally as the intrinsic velocity of mass-energy plus time at any point of computational space. In this view space itself is a field of mass-energy, a field of intrinsic spatial velocity with its density corresponding to the strength of gravitational fields present.

In this view the masses and other charges of particles are simply little units of c spacetime velocity that have broken away from the background and crystallized around their particle components. Thus they become free to move on their own relative to the background space. All particle masses and other forms of energy are actually forms of concentrated space velocity free to move against the background field of intrinsic spatial velocity. This theory is consistent with quantum field theory, which treats particles as excitations in space (Wikipedia, Quantum field theory).

If a unit volume of space contains strong mass vibrations the distance across it is further because the ups and downs of the vibrational waves must be traversed. This is completely equivalent to the curved spacetime model of general relativity because its curves could be compressed into vibrational waveforms in a flat Euclidean space and if they were stretched out again the resulting space would be curved.

In our model a mass is a field of minute vibrations in a flat Cartesian space centered on the massive particle(s) producing it. This is much easier to visualize than the curved spacetime of general relativity. The beauty of the velocity density vibrations model is its flat Cartesian space is very easy to understand and work with and presumably much simpler for the universe to compute.

In the vibrational space model the speed of light is the same for all observers everywhere in the universe. Light appears to move slower across vibrational space because actual distances across it including the ups and downs of the vibrations are much greater than they appear. Thus the actual distance traversed by light up and down the vibrations is such that light always does travel at c everywhere in the universe even in gravitational fields.

In the standard interpretation of general relativity this is due to the curvature of space in gravitational fields. Thus light beams are actually always traveling at c but when they have to traverse the curvature of space the distance traveled is greater than it appears so light appears to move slower than c to a remote observer. We don't directly see the curvature of space produced by gravitational fields because light travels along it but the actual distance is greater than the nominal distance between two points in a gravitational field. Vibrational density spaces model gravitational fields as flat Cartesian spaces as we actually see them rather than the curved spaces general relativity uses and this is another significant advantage of Universal Reality.

Thus we actually see the sun very slightly larger than we would if it didn't curve space near it and it's actually slightly farther away than its nominal distance measured along the curved space of its gravitational field. This effect is proportional to gravitational strength, which is why light appears to slow to zero speed as it nears a black hole and images of things pile up at the event horizon and fade out as they cross it (Wise, ?).

The STc Principle and the equivalence of gravitational mass-energy and spatial velocity, the METc Principle, are the two keys to understanding general relativity. And our understanding is greatly improved by replacing the spacetime curvature model with an equivalent intrinsic velocity density vibrational space model.

Even though time travels at different rates depending on different relativistic conditions, there is a hypothetical standard clock time rate for the universe, which is the maximum rate clock time can flow. This is the time rate of stationary clocks in deep space far from any gravitational field though there is nowhere this is strictly true.

This is the clock time rate of empty space, which is likely related to the intrinsic velocity of the zero-point energy of the quantum vacuum (Wikipedia, Zero-point energy). Since all forms of mass-energy produce gravitational effects the zero-point energy also produces a gravitational effect. And since all gravitational fields are equivalent to fields of intrinsic velocity, and the amount of intrinsic velocity sets the balance of space and time velocities there should be a relationship between the values of c, G (the gravitational constant), the zero-point energy, and perhaps the total mass-energy of the universe.

Thus the clock time rate of time in empty zero-point energy space is the standard baseline clock time rate of the universe. It would be the maximum possible clock time rate and all other clock time rates would be slower proportional to the presence of additional spatial velocity. Nevertheless all observers measure their own proper time rates as the speed of light on their own comoving clocks even if they are in a gravitational potential or moving through space as explained above.

ACTUAL VERSUS OBSERVATIONAL TIME DILATION

One of the important aspects of relativistic time is the difference between what can be called actual and observational time dilation. Relative spatial motion always produces time dilation but whether that time dilation is actual or just observational depends on what it's actually relative to. If the motion is relative to an actual world line through the background fabric of spacetime the time dilation is actual. Actual means that all observers agree on the amount of time dilation and the difference in elapsed time produced is permanent.

On the other hand two observers moving relative to each other will both observe the other's clock running slower by the same amount. The time dilation is reciprocal and vanishes with no permanent effect as soon as the relative motion stops. In this case the time dilation is merely observational.

In all cases of relative motion there is always some relative motion with respect to the spacetime background by one or both of the observers. But only the amount of relative motion with respect to the actual spacetime background produces actual time dilation. All the rest of the relative motion produces only observational time dilation.

Relativity itself has a hard time explaining the difference between actual and observational relativistic effects, as the founding principle of relativity is that all frames are relative and none are privileged or absolute. This is a major blind spot of relativity that is easily solved by a computational approach to reality.

The only way the difference can be explained is if there is an absolute spacetime background roughly aligned with the aggregate mass of the universe with respect to which the spatial velocity that produces actual time dilation is relative. In a computational universe this is simply the computational space in which the dimensionality of the universe is actually computed.

This approach has the added benefit of explaining what actual rotation is relative to and solving the problem of Newton's bucket; what actual rotation is relative to. This is explained in the next chapter.

Thus the takeaway is that for the observed effects of relativity including time dilation to actually make sense there has to be an absolute background space with respect to which the spatial motion that produces actual relativistic effects is relative. Combined with present moment time this gives us a universal background spacetime with respect to which all actual relativistic effects occur.

The equations of relativity are the same for both actual and observational effects because they are both an effect of apparent motion relative to the frame of an observer. The only difference is that actual relativistic effects produce permanent effects that all observers agree upon, while observational relativistic effects are seen differently by different observers and vanish when the relative motion ceases.

TWO KINDS OF TIME

We all directly experience the fact that there is a present moment through which time flows, and we know from relativity that time can flow at different rates through the present moment. The only way this most fundamental experience of our existence makes sense is if there are actually two different kinds of time. As strange as it may originally seem there is simply no way around this obvious fact.

There must be two different kinds of time, the time of the present moment and clock time which runs at different relativistic rates within a single universal present moment. The time of the present moment is universal and absolute and common to everything in the universe. And there is clock time, which flows through the universal present moment at different rates depending on the local presence of spatial velocity. Clock time flows through the current present moment at different rates but the universal present moment is common to all observers throughout the entire universe.

The fact there are two kinds of time is conclusively demonstrated by the established fact that relativistic observers always reunite in the exact same present moment even if their clocks read different clock times.

There is no doubt at all that there are two kinds of time and it's totally amazing that no one had recognized this obvious truth until I first pointed it out (Owen, 2007, 2009). This serves as an excellent example of the blindness of science to obvious facts that somehow just don't register in current memes even when perfectly consistent with established science.

Obviously this proposal is controversial and requires good evidence to be taken seriously. However it's pretty straightforward to demonstrate. First, to prove clock time and the present moment are two separate kinds of time we need only demonstrate there is a single universal present moment within which clock times vary since it's already an experimentally proven and widely used fact that clock times do run at different rates according to relativistic conditions. Thus if we can demonstrate this occurs in a common universal present moment, then clock time and the present moment must indeed be two different kinds of time.

The existence of a common present moment throughout the universe is not a new or strange idea. It was the standard accepted view of time throughout history until the advent of relativity. Clock time was thought to be flowing at the same rate throughout the universe and the present moment was thought to be the common universal present moment of *clock time* rather than a separate kind of time. The present moment was the current reading of a universal clock, which ran at the same rate throughout the universe. Thus there was only a single kind of time, clock time, and the present moment was the present moment of this universal clock time.

But with the advent of relativity it became clear that clock time didn't flow at the same rate everywhere so there couldn't be a universal *clock time* that was the same everywhere. So because time was still considered a single entity the newly variable clock time was still thought to be the only kind of time and the very notion of a universal present moment inconsistent with this view was wrongly discarded. The notion of two separate kinds of time to reconcile clock time with the present moment doesn't seem to have occurred to anyone until now.

Even though the present moment is a central experience of our existence scientists post relativity couldn't come up with any explanation of what the present moment was, and it was either ignored or even denied and replaced with truly outlandish theories such as block time in which all times exist simultaneously and there is no special present moment [2].

The main reason for the current ignorance of the present moment in modern physics is most likely that it has no obvious measure and physicists have the unfortunate habit of ignoring or denying the existence of anything without measure even though neither consciousness, nor existence, nor the present moment have measure and they are our three most important and fundamental experiences of reality.

So the important insight of this new understanding is to retain a universal present moment and recognize that relativistic clock times run at different rates within it. This is a very simple and reasonable insight in accord with our direct experience of reality, totally consistent with relativity, and it has profound consequences in our understanding of reality. And it also turns out this concept of two kinds of time is even an implicit though totally unrecognized principle of relativity itself without which relativity doesn't even make sense.

Every comparison of different clock times in relativity only makes sense if there is a common present moment in which the comparison takes place. There must be a common present moment that serves as a common background reference for clock times to be compared. Thus a common universal present moment in which *relativistic* comparisons can be made and shared is a hidden and completely unrecognized assumption of relativity used by physicists all the time but which many actively deny! This is one of the great blind spots of modern science.

For example if two space travelers with different clock times were really in each other's pasts and futures they would be completely unable to compare their clocks. They can compare the different readings of their clocks only because they and their clocks are both in the same present moment. Their present moment is the same but their clock times are different, thus it's clear there must be two different kinds of time.

When one of two twins embarks on a relativistic space journey they part in a common present moment. They then each continuously experience their separate existences in a present moment throughout the duration of their separation. And when they meet again they always meet in a present moment common to both even though their clocks now read different times. So there is no reason whatsoever to think the present moment they both experienced during the entire duration of the separation was not the same present moment for both. Their ages and clock times are now much different but their present moments never lapsed and must have always been the same even though their clocks were running at different rates.

Thus it's reasonable to assume, with no evidence at all to the contrary, that there is a single common universal present moment throughout the entire universe, and to assume that every observer in the universe is always in the same current universal present moment as everything else. Thus all that exists, the entire universe, exists in the single common current universal present moment.

This is really quite obvious, but it has been by far the most contentious aspect of the theory of Universal Reality with various arguments being raised against it. Some of these arguments have been based on a specific misunderstanding of the theory, that it violates the relativity of simultaneity in which different observers can have different valid observations of whether two events occur at the same *clock time* or not (Wikipedia, Relativity of simultaneity).

But Universal Reality accepts and incorporates all the equations of relativity including those of the relativity of simultaneity. The relativity of simultaneity correctly describes the behavior of *clock time*, but says nothing about *present moment time*. Non-simultaneous clock times are always compared in the same current present moment so this argument has no bearing on whether there are two kinds of time and can be disregarded.

For some very strange reason the existence of two separate kinds of time is beyond the comprehension of many otherwise intelligent people. Even though it's really quite a simple, logical and straightforward idea many people have great difficulty wrapping their heads around it.

SOME THOUGHT EXPERIMENTS

There are a number of useful thought experiments that more clearly demonstrate a universal present moment. To prove our point we must demonstrate both that there is a common universal present moment for all observers stationary with respect to each other and in motion or acceleration with respect to each other.

We already know that whenever any two observers are spatially collocated they both experience the same common present moment because they can communicate more or less instantaneously to confirm this. And this is true whether they are moving or stationary with respect to each other.

Consider first a universe completely filled with stationary observers packed tightly together like sardines. We already know every one of them is continuously in the same present moment as the adjacent observers on all sides and this is true for all observers in the universe even if all their clocks are running at different rates in different gravitational potentials. Therefore every observer across the entire universe must be in the same common present moment and this present moment must be universal.

This is a simple proof that there must be a single common present moment throughout the universe that holds whether the observers are stationary or in relative motion because it also holds if the observers are moving relative to each other and just happen to transiently assume the packed sardine configuration when the experiment is performed.

This proof also holds whether or not there are gravitational fields involved. The clocks of observers in gravitational fields will be running slower and across the entire universe all clocks will be running at varying rates in the same universal present moment and this will be agreed by all observers as each confirms their existence in the same current present moment with all those adjacent.

Now the counter argument might be raised that there is some slight difference in present moments across the entire universe too small to be noticed by spatially adjacent observers. But what could a difference in present moment even mean? The present moment most certainly doesn't correspond to differences in clock times because we already know different clock time rates exist in the same present moment. So what would such a difference in present moments amount to? What would cause it and why would that difference be magically erased when separated space travelers meet with different clock times? If their present moment times were different during their separation how could they become the same again when they met? What would be the mysterious mechanism involved? It wouldn't make sense.

Now consider another thought experiment involving two space travelers who part with synchronized clocks, accelerate at different velocities in different directions and periodically redirect to cross paths. We know they are in the same present moment whenever they cross paths so take this example to the extreme and assume they accelerate with enormous and varying velocities but turn and cross paths more and more frequently until the interval between meetings approaches zero. Now they are crossing paths every minute fraction of a second. And every time they cross paths they both confirm they share the same present moment, and no matter how they change their accelerations and how much their clock times vary this is always true.

So it seems outlandish to assume that somehow in the minute fractions of a second they are separated their present moments somehow become different. Their clocks can run at very different rates and read differently every time they cross paths but this is always happening in the same shared present moment. And in fact they both can confirm to each other their clocks are running at different rates every time they cross paths in the same present moment.

This is also confirmed by ground communications with the International Space Station. Time aboard the ISS progresses at a measurably slower rate than on earth because of the velocity it's traveling along its world line (Wikipedia, Time dilation). Yet the ISS is in continual contact with the earth at all times. The ISS and earth are both in the same shared present moment at all times even as their clock times run at different rates. True there is a slight communication delay but this delay occurs in the common present moment.

Now imagine the ISS accelerating enormously but maintaining the same circular path around the earth. Clock time and all physical processes aboard the ISS will begin running perceptibly slower from the point of view of earth observers and vice versa, and this can be confirmed by continual mutual communication. The slight time delay in communications remains the same confirming it's irrelevant. Observers on both earth and the ISS would now see each other's clocks and people moving in slow motion but they would both continually communicate this fact back and forth in the same shared present moment even though each would hear the other speaking slower.

Their times would certainly seem very strange to each other but since they remain relatively close throughout they both would be able to continually observe this mutual strangeness in the same present moment. They would both continually observe each other's clock times running at wildly different rates within the same shared present moment.

So it's clearly possible for two observers to actually watch each other's clocks run at different rates in the same present moment in some cases. This can only be true if the present moment and clock time are two different kinds of time. Science must be based on observation and the present moment and two different kinds of time are clearly observable facts.

In light of these thought experiments it seems undeniable that there is a single present moment common to the entire universe that is completely different than clock time, and therefore there are two completely different kinds of time.

THE PRESENT MOMENT

It's important to clearly understand what is meant by the *same* present moment. There isn't a single present moment that stays the same over all time as clock time flows through it at different rates. Instead there's a separate present moment kind of time that also progresses but at the same rate throughout the universe. The current present moment right now isn't the same as the previous present moment. For things to happen there must be a current present moment that is not the same as the previous current present moment.

The present moment time is the successive present moments of the active process of happening in the observable universe, which continually manifests as a universal present moment in which everything exists, and in which the state of the universe is continually being recomputed. This theory of two kinds of time is completely consistent with science including relativity and with personal experience and informed common sense.

Happening can be considered as the processor that continually computes the current data state of the entire universe. Since all the data of the universe exists simultaneously within the processor of happening the *presence* of existence manifests as a universal current present moment in which everything happens. This current present moment is universal and common to all observers no matter how fast their clocks are running.

Thus happening is the source of the fixed flow rate of present moment time throughout the entire universe. The time of the present moment flows at the same universal rate throughout the universe, but within that universal flow the different relativistic flow rates of clock time are computed locally on the basis of the STc Principle.

However it's not easy to pin present moment time down because it has no intrinsic metric since all clock time metrics are computed within it. Thus present moment time is prior to the computation of any observable dimensionality of time. Happening computes all dimensionality including clock time metrics so it has no observable measure of its own.

If present moment time and computational space are the source of all dimensional space and time metrics they can have no observable metrics of their own. Thus they are pre-dimensional in the same sense that a numeric representation of dimensional data in a computer program is not itself a dimensional structure. Thus a computational universe is intrinsically non-local and events across the entire universe can be computed simultaneously irrespective of the finite speed of light. This has important relevance to the apparent problem of quantum non-locality as explained in the chapter on Computing Quantum Reality.

Thus there is a present moment time that progresses with happening but no intrinsic associated metric other than the various clock time rates that it locally computes. So two distinct kinds of time do exist and both progress, but only clock time has a measurable metric associated with it. Thus present moment time can only be measured in terms of the clock times it produces.

There is also another way to get a sense of the dimensionality of present moment time in terms of its effect on cosmological geometry as explained in the section on the Hyperspherical Universe. In fact the concept of a separate present moment time is doubly important because it immediately nails down the previously uncertain geometry of the observable universe.

P-TIME

If we call the time of the present moment P-time, then every actually occurring event takes place simultaneously for all observers at the same P-time throughout the entire universe. In other words if an event takes place in the present moment for one observer, it also takes place at the exact same P-time for all observers. This will always be true no matter how different the clock times and clock rates of various relativistic observers are. This universal simultaneity of P-time is what allows relativistic observers to compare their different clock times in the same present moment.

This P-time simultaneity has nothing to do with the relativity of simultaneity of *clock time*. Different observers can still have different views of the clock time simultaneity of events due to the finite speed of light between observers and events. However the entire observable universe, including its different clock times, is continually recomputed in the same universal P-time moment.

There is another good argument for a universal P-time, but we first need a couple of basic definitions from relativity. Relativity defines *proper time* as the clock time reading of an observer's own comoving clock, the clock on his wrist or wall. *Coordinate time* is the time an observer sees on another observer's clock and is a measure of the intrinsic rate of processes associated with that other observer from the perspective of the first observer. An observer's measurement of another observer's coordinate time is always by comparison to his own proper clock time, and that measurement, like all observations, is always made in the first observer's present moment.

The proper times of all clocks are continuous; there are never any gaps in the flow of their P-time. There is never any moment that proper time is discontinuous during the separation of observers or at any time for that matter. Thus there must always be a one-to-one correspondence of proper clock times for any two observers. There must always be some proper time reading on one observer's clock for every proper time reading on the other's clock whether they are different or not.

This gives us an additional means of testing the universal present moment theory. Every observer can specify his progressive P-times in terms of his own proper clock time readings. There is always a proper time clock reading for every current present moment since both clock time and P-time are continuous. For example, observer A can say he was tying his shoes in his present moment when his proper time clock read 12:00 AM. Every observer's proper time clock reading serves to uniquely identify his own current present moment P-time at that time, and this is true of all observers. For every current present moment of every process in the universe there was always a corresponding proper time reading that can be used to denote it from the perspective of any observer whether or not those proper times are identical.

Therefore if there is a single universal P-time that flows at the same rate for all observers so they all remain in the same common universal present moment, then there must also be a unique one-to-one relationship between the proper times of all observers even when they differ. There must be one and only one proper time on every observer's own clock that corresponds to every single P-time they all shared when their clocks read those proper times. If there isn't then the universal present moment theory is falsified, but if there is it's confirmed.

Simply stated, for every present moment proper time of any one observer every other observer must have been doing something at that exact same present moment at their own corresponding proper times, no matter how differently their own proper time rates may have been flowing relative to one other.

In many cases this proper time correspondence can be calculated and confirmed. If the space traveling twins exchange flight plans before they separate the one-to-one correspondence of their proper times throughout their separation can be calculated. Each will know the complete relativistic history of the other and thus know exactly how much proper time has elapsed on the other's clock for any proper time on its own clock. Each observer will know what the other's clock is reading at every moment on his own clock. To be absolutely clear this *calculation* of the current *proper time* of the other twin is not the *coordinate time* that would be *observed* on the other twin's clock. It's the *proper time* of the other twin's clock, which is not generally observable but which can be calculated.

Thus it's always possible for any observer who has knowledge of the relativistic circumstances of any other observer to calculate what proper time reading of that other observer correlates to each of his own proper time readings. There is always a one-to-one correspondence of proper time readings between any two observers that tells them what proper time of one corresponded to the proper time of the other in their common present moment even when they are separated in different relativistic circumstances with different clock time rates.

However in the general case of any two observers in the universe, where an initial proper time correspondence can't be determined or the relativistic history of the other observer is not known, it may be impossible to calculate the current proper time correspondence even though it's certain one must exist. However if any observer in the universe can determine the relativistic variables of any other observer as well as his own he can calculate the proper time rate of that observer relative to his own proper time rate and confirm the existence of a common shared P-time.

Thus it's easy to show there will always be a one-to-one proper time correspondence between any two observers in the universe, and this is all that's necessary to conclusively demonstrate a universal P-time present moment common to all observers. It's sufficient to note that time is continuous for all observers thus every observer in the universe must be doing something at every proper time moment of any other observer's clock. Whatever is being done will be done in the exact same common universal present moment of all existence in the universe because the present moment is the only time that anything can occur because it's the only moment that exists and the only locus of reality.

The clock times of different observers can flow at different rates through this common present moment, but there is always a one-to-one correspondence between the proper times of any two observers throughout the entire universe. This is consistent with Universal Reality's proposal that the common universal present moment of existence is all that exists, and is the current moment of happening in which all the computations of the universe take place. It is the current universal P-time tick of the entire computational universe.

Therefore the proper times of observers can be used to notate the passage of P-time, and their correspondences, when they can be determined, can be used to establish identical P-times among observers even though P-time has no intrinsic metric of its own. And if all else fails any two observers can simply communicate their current relativistic conditions and proper times to enable their P-time simultaneity to be calculated.

THE ARROW OF TIME

One of the perennial mysteries of science has been the source of the arrow of time; the fact that time continuously flows in the forward (by convention) direction. Many scientists have sought the source of the arrow in vain, often mistakenly attributing it to entropy (see the section on 'Entropy and Time' for why this isn't correct), but the explanation is quite straightforward and a simple consequence of the STc Principle.

The STc Principle states that everything continuously travels at the speed of light through time on its own clock. Therefore it's quite clear that time must be experienced as flowing in a single direction by every observer and that every observer's own clock will always appear to have its hands moving at the speed of light in the same direction as all others.

Thus the STc Principle itself is the actual source of the arrow of time, and puts the arrow of time on a firm scientific basis. The arrow of time is a direct implication of the theory of relativity; another completely unrecognized fact of modern physics, and even actively denied by scientists who don't understand it.

CONFIRMING A PRESENT MOMENT

Amazingly the necessity of a privileged present moment of time distinct from all other moments of time is another direct consequence of the STc Principle, again completely unrecognized by most physicists. In fact many physicists continue to deny the existence of a present moment mistakenly believing it's inconsistent with relativity. It is not. It is actually *required* by relativity and a direct consequence of the STc Principle that underlies special relativity.

By the STc Principle everything continuously travels through spacetime at the speed of light. This means everything must always be at one and only one point in time, and that point is obviously the current present moment of its actual existence. Thus the STc Principle requires the existence of a privileged present moment for all observers that is the current time of their existence, and the *only* time they are actually at, the time that defines their now.

Thus relativity itself absolutely requires a present moment that progresses in clock time and conclusively falsifies the nonsensical 'block universe' hypothesis in which all times exist at 'once' as a single static structure (Price, 1996, 12-13, 14, 15-16), (Wikipedia, Eternalism (philosophy of time)) [2].

This present moment required by relativity is the same present moment we have already identified as the presence of existence that intrinsically manifests as a single universal present moment in which everything exists. Thus Universal Reality is clearly consistent with and confirms a proper understanding of relativity in this respect.

Because existence exists it must have a presence and its presence manifests as a universal present moment in which everything exists. The universal present moment is simply the presence of existence and encompasses all that exists and there is no before or after it that actually exists. The presence of existence is the source of the present moment that is now confirmed by the STc Principle to be consistent with relativity.

Physicists who deny the existence of the present moment should remember it's the most fundamental and persistent of all observations, and that the role of science is to explain observations, never to deny them.

THE HYPERSPHERICAL UNIVERSE

Relativity has revealed that we live in a 4-dimensional universe consisting of 3 dimensions of space and 1 dimension of clock time and this is clearly correct. However relativity itself hasn't been able to discover the actual geometry of our universe because it hasn't recognized the present moment as a second kind of time. The unfortunate result is science has no clear picture of the overall geometry of the universe and tends to fall back on clearly inaccurate expanding tube like images with flat surfaces and edges (Wikipedia, Metric expansion of space, #Topology of expanding space).

Modern science is positively schizophrenic when it comes to the notion of a universal time. First physicists adamantly deny the concept of a present moment and the notion of a common universal time, and in the next breath they tell us the universe is 13.8 billion years old and that's true for every observer in the universe. Then they engage in all sorts of genuflections to try to reconcile these clearly contradictory views.

The obvious cosmological geometry of the universe is a 4-dimensional hypersphere where the 3 surface dimensions are our 3 dimensions of space and the radial dimension is the time dimension from the surface back to the center point of the big bang.

However this doesn't work if the radial dimension is clock time because clock time can run at different rates within the surface and there could be no single consistent radial time dimension. This is why physicists haven't been able to discover the true hyperspherical geometry of the universe.

However if we recognize the separate time of the present moment and take the radial dimension as P-time instead of clock time the hypersphere works quite well because the P-time rate is fixed across all regions of the surface and we get a single consistent P-time radial dimension for the universe. The obvious 4-dimensional hyperspherical geometry now makes perfect sense.

Everything that exists, the entire universe, exists in the current present moment surface. The past no longer exists and all the past onion-like layers of the hypersphere where the surface used to exist correspond to the history of the universe over past P-time back to its origin at the point of the big bang at the center.

Thus at each P-time tick the entire surface of the actually existent universe is recomputed and a new very slightly larger surface in the next present moment is created. Thus the 3-dimensional spatial surface of the universe continually expands as P-time progresses.

As the progression of P-time continually expands the surface local clock time rates are computed across the surface depending on the presence of spatial velocity according to the STc Principle. The 3-dimensional surface is the current present moment fabric of space and consists of a spacetime velocity at every point equal to the speed of light c. Any intrinsic velocity of a gravitational field at any point automatically reduces the velocity of time at that point so their vector sum is always c.

Thus the universe takes the form of a closed finite 3-dimensional hyperspherical surface in the current present moment with positive curvature and no edges at the largest scale. It cannot be infinite because nothing actual can be infinite because infinity is not an actual state or fixed number but a never-ending *process* of continual addition. Infinity is a useful mathematical concept but nothing actual can be infinite.

Nor is there any reason to believe the universe is not a closed continuous surface and has edges. How could the point universe of the big bang develop edges as it inflated? That would tear it apart and it's clearly nonsensical. Traveling in a straight line in any direction across the universe one would theoretically eventually end up at approximately the same place ignoring any local curvatures of space just as one does by circumnavigating the earth ignoring the mountains and valleys.

If P-time is the radial time dimension of a hypersphere then the circumference of the spatial surface of the universe will be a function of its P-time radius, and measurements of the curvature of space should provide a measure of its radius and the P-time age of the universe. Current measurements suggest that 3-dimensional space is fairly flat within its observable volume but a hypersphere is not ruled out (Wikipedia, Flatness problem). A hypersphere also makes sense from the perspective of general relativity, as the mass-energy content of the universe should curve it in on itself at the largest scales.

This hyperspherical geometry should be subject to experimental confirmation since the curvature of space is measurable (Wikipedia, Shape of the universe). It should turn out to have a very small positive curvature. But even if it doesn't that raises doubts but doesn't necessarily falsify the hyperspherical geometry since if the hypersphere is not perfect it could be closed and finite and still contain some areas with greater or lesser or even possibly negative curvature.

Only a small volume of the cosmic hypersphere is visible from any location within it since its spatial surface is uniformly expanding at a rate that exceeds the speed of light beyond a distance called the particle horizon. Because space itself is expanding away from us faster than the speed of light beyond the particle horizon, light can never reach us from beyond there and that area of the universe isn't visible to us. Likewise we are not visible from points beyond our particle horizon because we are beyond the particle horizons of those points.

However the entire current P-time surface of the hypersphere including all regions beyond the particle horizon is the whole actual universe since the entire surface is the current present moment of P-time in which the entire universe exists and is recomputed. The entire surface of the hypersphere, the entire universe, is in the same P-time present moment all around its surface, irrespective of particle horizons, and irrespective of the various local rates of clock time.

Note that the particle horizon is an observational as opposed to an actual relativistic effect. Beyond the particle horizon nothing is actually moving faster than the speed of light with respect to the background fabric of spacetime. Processes evolve normally just as they do in our area of the universe. The absolute dimensional background of computational space in which all processes are computed extends around the whole surface of the hypersphere with no interruptions or anomalies. It's only when processes near some observer's particle horizon are observed from afar that anomalies appear to exist. But all such anomalies are anomalies of clock time observations and have nothing to do with P-time, which continues to recompute the entire surface with each universal P-time tick.

INFLATION & THE HUBBLE EXPANSION

The hypersphere is a neat and elegant model of the universe but it initially seems to have two problems. First the rate of the P-time radius expanding the surface is presumably constant but the expansion of the surface seems to have varied greatly over the history of the universe and the current Hubble expansion of the surface appears to be accelerating (Wikipedia, Accelerating expansion of the universe). If the rate of increase of the radius is constant it seems the expansion of the surface should also be constant.

And second if the radius of the hypersphere was only the 13.8 billion years of its clock time age back to the big bang the size of the universe would have to be much smaller and its curvature would have to be very much greater than current measurements suggest.

The solution to these apparent problems is straightforward. Science measures the expansion of the surface in *clock time* but the extension of the radial dimension of the hypersphere is in *P-time* and their rates are not proportional because P-time computes various clock time rates depending on the presence of spatial velocity. For example by the STc Principle if there were a time when the spatial velocity in the universe was much greater than it is now the overall clock time rate of the universe would have been much slower.

In fact there was such a time called inflation. The inflationary period of the universe was an enormous exponential expansion of the universe in the first slight fraction of a second after the big bang (Wikipedia, Inflation (cosmology)). There is considerable evidence for inflation and the theory is widely accepted.

Inflation was an enormous explosion of the spatial velocity of the entire universe. Thus by the STc Principle there must also have been an equally enormous slowing of clock time throughout the entire universe. This means that the overall clock time rate of the universe slowed to almost nothing and clock time barely passed at all during inflation because the same rate of P-time happening was busy computing the spatial expansion instead.

Thus from our current look back rate of much faster clock time, inflation seems to have occurred almost instantaneously in clock time while presumably P-time would have been running at its standard rate. Thus enormously more P-time would have passed than clock time, which means the P-time age of the universe is enormously greater than its apparent clock time age of 13.8 billion years.

This means that the actual radius of the hypersphere is much greater than 13.8 billion light years and its circumference will also be much greater. This immediately allows the curvature of the hypersphere to be consistent with current measurements of the curvature of the universe and the apparent problems with the hyperspherical model can be resolved.

So clearly the changing rates of expansion of the universe in clock time over its history can be decoupled from the presumably uniform rate of extension of its P-time radius. Thus the changes in the rate of Hubble expansion of the universe over its history including its current apparently accelerating rate are consistent with the hypersphere model.

After the initial period of exponential inflation, the expansion of the 3-dimensional space of the universe seems to have quickly decelerated to a much slower rate before beginning to gradually accelerate again. The expansion appears to be still accelerating. This expansion of the universe is called the Hubble expansion after its discoverer, Edwin Hubble (Wikipedia, Hubble's law).

Even though the expansion is imperceptible at local scales, over intergalactic distances it adds up to produce particle horizons equidistant from every point beyond which the expansion exceeds the speed of light and nothing is visible.

So the Hubble expansion is an extraordinarily slow expansion of the fabric of space. At any given point of space it's completely imperceptible and only becomes apparent over intergalactic distances. So any minute changes in expansion rate in the spacetime fabric over time are very easy to reconcile with the presumably constant rate of the P-time extension of the radius of the hypersphere by very slight changes in the overall spatial velocity content of the universe very imperceptibly reducing the overall clock time rate.

Any expansion in the fabric of space is spatial motion and thus slows the local clock time rate. The Hubble expansion is a fairly uniform expansion of the entire surface of the hypersphere and thus the clock time rate of the entire universe is slowed. However that slowing would be locally imperceptible since it occurs point by point across the universe where the expansion is likewise far below the level of possible measurement. The expansion only shows up at intergalactic scales as red shifts as does the slowing of clock time in the greater time it takes light to cross intergalactic distances across expanding space and the stretching of its frequency which is what produces red shifts.

So the continual P-time extension of the radius of the hypersphere does produce the Hubble expansion, but when it's measured in clock time the Hubble expansion can vary because the expansion automatically produces a concurrent slowing of clock time.

It should also be pointed out that though the Hubble expansion appears to be accelerating it's not clear how accurately we even know the expansion rate over time. Due to the finite speed of light we are only able to observe the universe as surfaces of fixed distance and time. So we have no direct observations at all the other distances for any given time, and all the other times for any given distance. This gives us only a minute sample set of the rates of expansion over the observable history of the universe.

In particular we have no idea at all of what the *current* expansion rate of the universe really is since it's totally unobservable, and will remain so until the light from the standard candles used to measure it begins to arrive which will be far into the future.

Standard candles are objects such as Type 1a supernovae, which all have approximately the same intrinsic brightness and thus allow very accurate estimates of their distance by their relative brightness. Their apparent brightness gives an accurate estimate of distance, and their red shift gives a measure of the recession velocity of space at that distance. However another problem is that type 1a supernovae are quite rare and transient and widely dispersed so the sample size upon which results are based is extremely small.

However we do have some inkling of expansion rates over time, as the expansion rates of any past time distance surface are generally consistent across the surface. Also the red shifts that are used to measure distances to the standard candles are caused not by the recession velocities of the objects emitting the light as often thought but by the cumulative stretching of space over the entire distance between us and the object. So red shifts do give us some sense of the expansion rates of the spaces and times between us and the standard candles especially when red shifts from different past times are correlated.

So an accelerating Hubble expansion is a reasonable conclusion, even though again the closer in time and space the standard candles are to us the less precise the expansion rates of the universe there are. In the time scale of human history we have no information whatsoever because the nearest standard candle is many light years distant.

So both inflation and the varying expansion rates of space over time are completely consistent with the hyperspherical geometry of the universe when we take its radius as a P-time dimension rather than a clock time dimension. Thus the P-time radius of the universe is going to be much larger than its apparent clock time radius of 13.8 billion years, and its nearly flat observed curvature is completely consistent with that.

SEEING ALL 4 DIMENSIONS

We can actually confirm the 4-dimensional hyperspherical geometry of the universe visually because we can actually see it with our own two eyes. There has been much discussion about how to visualize the 4-dimensions of spacetime, of how to see the time dimension just as we see the 3 spatial dimensions. However the fact is we already see all 4 dimensions of the universe all the time laid out clearly before our eyes.

We see down the time dimension into the past as distance in every direction from every point in our 3-dimensional space. This is called our light cone and it's our personal view of the 4-dimensional cosmic geometry of the universe from the singularity of our location in our present moment of spacetime.

We see all 4-dimensions but there's a catch because the light cone we see is only a slice through 4-dimensions rather than the entire 4-dimensional universe. We see the past only as it existed at certain distances, and we see surfaces of space only as they existed at particular times in the past. Thus we neither see all of space nor all of time, but only a slice through both centered on our singularity.

Our experience of the passage of time through the present moment is our direct experience of the fundamental process of the universe, the continual recomputation of the information state of the universe including the passage of clock time through the present moment. This manifests as the 4-dimensional hyperspherical spacetime we directly observe around us.

SINGULARITIES IN TIME

Our location in spacetime is a singularity in the sense that clock time continuously flows into existence through the point of our location and then out in all directions into the past. Thus only our own current location exists in the present moment on our own clock. Everything else in the universe is at some distance from us and thus exists at least slightly in the past relative to us from the perspective of our present moment. Thus every observer exists alone in his own clock time singularity from which he observes the rest of the universe as it was in the past.

Of course everything and all observers actually exist in the same universal present moment but that common existence is not directly observable due to the finite speed of light. Our actual experience of all other things and observers in our present moment is always a no longer existent past representation down the radial time dimension of the universe.

Clock time continuously flows in from non-existence through our singularity into the present moment. The future continuously becomes the present as the state of the universe is continuously recomputed at the point of our existence. But there is no actual future that we reach that then becomes the present. The present state of the universe is just continuously recomputed in the present moment and clock time is simply the local observational rate at which those computations happen.

Though only the present moment has reality an observer can think of clock time as continuously becoming into being at his singularity and then flowing out into the past into the distance along the radial time dimension in every direction. Everywhere we look in the universe we see the past receding from us from the back of the moving train of time into the distance along the radial time dimension of the universe.

Everywhere we look we look into the past receding from our eyes, and nowhere do we see the future approaching except in our imagination. Thus our singular location in space and time is the point of the continual creation of existence, and once created the universe flows out into the past in all directions away from us.

SPACE TRAVEL

Due to the STc Principle a clock moving through space will run slower than a clock at rest, the slowing depending on the spatial velocity along its world line relative to computational space. Traveling along a world line in space will always take less time on the traveler's clock than the clock of a stay at home observer.

This slowing of a clock traveling through space can be enormous as its velocity approaches the speed of light, a fact that makes interplanetary travel theoretically feasible, at least with respect to the time required. Calculations show that a trip from earth to the center of the galaxy at a constant 1g (the equivalent of earth's gravity) acceleration for half of the trip and a 1g deceleration for the other half would take only 42 years on the clocks of the travelers, though well over 42,000 years, a little over the distance in light years to the center of the galaxy, would pass on clocks back on earth or at the galactic center (Misner, Thorne & Wheeler, 1973).

Of course a propulsion system that could produce a constant 1g acceleration for 42 years is not currently available and the difficulty of detecting and avoiding any intervening objects at close to light speed is near impossible. Nevertheless time dilation does make interstellar travel a theoretical possibility. So alien civilizations, if they exist, could just as easily travel to earth as well. The time it would take on their clocks would be quite acceptable even though it would take a very long time by our earth clocks and by clocks back on the aliens' home planet.

TIME TRAVEL

There are many misconceptions about time travel, especially when the significance of a universal present moment is not understood. However when we understand that the present moment is the only actual locus of existence and clock time runs at different rates within it everything becomes clear.

Understanding time travel and its constraints is pretty simple if we just keep in mind these two basic principles. First everything always exists in the same universal present moment and can never leave it because all that exists does so within this universal present moment and there is nowhere outside of it go. And second clock time flows at different rates within this universal present moment based on the presence of spatial velocity. That's really all there is to time travel.

We are all already traveling through time at the speed of light on our own comoving clocks all the time. So we are all already time travelers in this respect. We can't not travel in time because the passing of time is precisely us traveling through time at the speed of light. This is a basic implication of relativity and the STc Principle. What we experience as the passage of time is us traveling in time at the speed of light.

If we are moving fast through space or are in the strong intrinsic spatial velocity of a gravitational field our spatial velocity is subtracted from our velocity in time and our clocks slow down. Thus we may travel through the present moment at different clock time rates depending on our relativistic conditions, either because we have different relative motions or are in different gravitational potentials.

Though we are all traveling through clock time at different rates we all stay in the common universal current present moment. There is no possibility of traveling out of the universal present moment because it's all that exists. The present moment is the only locus of reality and of the entire actual universe.

These are the actual limits on time travel. No going back to the past or forward to the future. The future doesn't exist so there is nowhere there to go. Likewise the past doesn't exist so there is also nowhere to go in the past either. And sorry no wormholes through time either (Thorne, 1994, p. 483). We all stay in the common universal present moment, but our clock times, and all associated physical processes including our aging, can progress at different rates within the common present moment.

So we certainly can travel in time at different rates in the forward direction of time's arrow. There is extensive observational proof of this and relativity describes it precisely. Space traveling twins can separate and meet up again with different ages, but this is not the same as actually traveling into the future or the past by either twin. Both stay in the same universal present moment at all times and can never leave it. One just ages faster than the other in that present moment. The notion of traveling to an actual past or future out of the present is simply impossible.

So unfortunately there is no going back in time to view dinosaurs, and also no going back in time to change things there that alter the present. Thus there are no possible time travel paradoxes. And as interesting as it would be, no arrival in the present of time travelers from the future. It's simply impossible because the future has never existed because it hasn't been computed and there is nowhere to arrive from. It doesn't exist until it's actually computed in the present moment and then it becomes the present.

We have all arrived in the present moment from the past, but we have never left the continually evolving universal present moment to do so. But it is true that the past a space traveler arrived from could be very far back in time by our clocks if his clock was running very much slower than ours. With the right interstellar flight plan he could have left Earth when Nero was emperor of Rome and arrived back here just yesterday not a lot older than when he left Rome. In the colloquial sense that could be called a person arriving in the present from the past even though he never actually left the present moment as the centuries passed.

And neither he nor we nor anyone could ever travel back to his or any actual past since only the present moment exists and the entire past including the Roman Empire is irretrievably gone.

So we can certainly arrive at the same location in the present moment from different original past times, and that could certainly be very interesting, but everyone is continually in the common current universal present moment during the entire duration of his or her lives and travels. Some lives could be very much longer than others according to other clocks but only if they lived at much slower rates.

Our ancient Roman space traveler could arrive back on earth today to meet his 60th generation grandson and catch up on 2000 years of missed Earth history. Again extremely interesting but at every second during those 2000 years he and the earth would have both existed in the same current present moment. Events on earth would have just been progressing at a much faster clock rate than aboard the Roman space ship.

It is also theoretically possible for you or I to travel to an arbitrary date in the future in the same colloquial sense by taking a space flight with the right velocity and slowing down our clock time. But this is just a matter of slowing our clock in the universal common present moment relative to the rate of clocks at our destination. No one ever leaves the common present moment but we could arrive there with much less elapsed time on our own clock. However this requires a very high velocity space flight or intense gravitational field.

Because our spacetime is very close to flat on earth and we have a very low velocity relative to the background there is no way anyone else's time could be running appreciably *faster* than our own and there are really no additional effects to consider on that basis.

Everything that exists always exists in the same universal common present moment at all times as it evolves but time travelers could certainly arrive in the present from deep in the past with first hand information and even photos and videos given the proper technology. We can only hope!

ENTROPY & TIME

Entropy is the tendency for the energy states in any isolated volume of space to reach equilibrium over time. For example in a completely insulated box objects at initially different temperatures will eventually all reach the same temperature. Thus presumably the entire universe will eventually reach an energy equilibrium in which no additional transfer of energy can occur and all processes will come to a halt (Wikipedia, Heat death of the universe).

This energy equilibrium is not perfect nor is it necessarily eternal due to random zero-point energy fluctuations in the quantum vacuum, which are not subject to entropy and continually affect the state of the universe of actual particles. But these effects are statistically extremely unlikely to produce any large-scale energy imbalances that affect the progression towards maximum entropy.

Because entropy appears to be a fundamental unidirectional process in time that seems irreversible some physicists have proposed that it's somehow the source of the arrow of time but this is not correct (Price, 1996, p. 22). We have already correctly identified the STc Principle as the source of the arrow of time, and more fundamentally the fact of the happening of existence, which continually computes the evolution of the universe, is the ultimate source of clock time and its arrow.

And entropy can't be the source of the arrow of time because it varies wildly from region to region. There are many areas of the universe in which entropy is decreasing due to incoming energy and there is certainly no reversal of the arrow of time in those areas. If entropy were responsible for the arrow of time it would have to be a universal aggregate effect rather than a local effect.

However there is no physical mechanism that could account for such a universal effect. For one thing entropy is entirely a *result* of physical processes rather than the *cause* of anything. And more importantly entropy states depend entirely on the current mix of fundamental forces at any location.

Entropy states are not fundamental, as usually assumed, because they depend on the spatial mix of prevailing forces. For example cosmic scale entropy states reverse if gravitation reverses, and at smaller scales entropy states depend on the distribution of the other three fundamental forces.

In an initially stable universe with only attractive gravitation the ultimate maximum entropy state will be a single black hole because all matter will eventually clump together. But in a universe with only repulsive gravitation the ultimate maximum entropy state will be a continually expanding universe in which all matter continues to fly apart forever. Thus entropy reverses if gravitation reverses.

In our expanding universe where there is an apparent mix of attractive and repulsive (dark energy) gravitation and that mix seems to be changing it's unclear what the final maximum entropy state will be.

Thus cosmological discussions of entropy are almost always flawed because they fail to recognize that entropy itself is not fundamental (Penrose, 2005, p. 690). What is fundamental is the force mix including the expansion of space that defines the measure of entropy. Entropy is meaningless without reference to the force mix it's relative to. Maximum entropy has to be redefined as a state of energy equilibrium *under the mix of prevailing forces*.

Thus entropy is not a fundamental principle as usually thought. It's entirely a *result* of the evolution of the actual fundamental computational principles. Like all emergent laws it describes reality but doesn't actually compute anything.

When the dependence of entropy states on force mix and distribution is understood it becomes clear entropy has no causal connection to time and is certainly not the source of the arrow of time.

CONCLUSION

The central experience of our existence is our consciousness in a present moment of time within which happening occurs and clock time passes at the speed of light. The present moment is universal and is simply the presence of reality. Thus all observers in the universe exist within the same universal present moment in which the entire universe exists and there is not even nothing outside, before or after.

It's clear from relativity that clock time passes at different rates depending on the presence of spatial velocity within this shared universal present moment. It is also clear from relativity that all observers in the universe continually travel forward in clock time at the speed of light as measured by their own clocks. And it's clear that all observers see the 4th dimension of past clock time as distance in every direction from their location in the 3-dimensional space of the universe.

All these aspects of time can be directly realized in our experience. If we turn our attention to the passage of happening and clock time through the present moment we find our consciousness of this process is indeed the fundamental experience of our existence. We just need to realize that this experience is us and everything around us traveling at the speed of light through the 4th dimension of time even while we sit on our sofas. We are all surfing the 3-dimensional surface of our expanding hypersphere at the speed of light as we ride the evolving wave of existence.

And with the assistance of science we can directly experience the fact that clock times passes at different rates depending on the presence of spatial velocity. If we observe the half-lives of decaying particles moving at relativistic rates, the speed of our clocks on earth relative to those traveling in space, or even by directly comparing our clocks to those returning from space flights we can directly experience this. They can all be directly realized in relativistic circumstances in our daily lives. Even magnetism is our direct experience of the effect of electric charges moving with relativistic velocities as explained in the chapter on Relativity And Electromagnetism.

We can also directly experience and realize the continual computational creation of the information state of the present as a process that occurs only within the happening of the present moment, thus realizing the non-existence of the future. We can also directly realize the non-existence of the past even though we observe it as distance in every direction because we are observing that and everything else in the universal present moment of all existence as its light arrives in our eyes.

Thus we immediately realize the impossibility of time travel in the sense of traveling out of the present moment. The present moment is all that exists and where everything exists and happens. We can see down the past dimension of time only because of the finite speed of light. We are not actually observing the past, which is nonexistent, but the light trace of the past in the present moment.

The very fabric of space itself consists of the fixed speed of light velocity of spacetime at every point. That velocity can be either velocity in space or velocity in time but is always equal to the speed of light c. Thus whenever velocity in space increases velocity in time decreases and this STc Principle is the key to understanding both time and relativity.

The two fundamental aspects of time are the present moment and happening. The present moment is simply the manifestation of the presence of reality. And happening is the continual recomputation of the data that constitutes the universe.

The rate of happening is what we call clock time. If we think of the universe as a computational system the processor that computes the universe has a fixed cycle rate that is the direct source of the fixed spacetime speed of light velocity of every point and process in the universe. When velocity in space increases there are fewer processor cycles left over to compute velocity in time and time dilation occurs. This is the computational source of the STc Principle and most of the effects of relativity.

Putting thus all together we arrive at a deep and comprehensive understanding of the mystery of time, what it is and how it works, and its major implications for the nature of reality itself.

COMPUTING RELATIVITY

MASS VIBRATIONS & GRAVITATION

We begin this chapter with an outline of the complete theory of vibrational mass and how it explains the relativistic effects of gravitation in an equivalent but much easier to understand manner than general relativity. This model is superior to the curved space model of general relativity because it models spacetime as a flat Cartesian space with fields of intrinsic velocity density, which is how we actually see it. It also provides an explanatory mechanism for how the presence of mass affects space that is missing from general relativity, which never explains why or how mass actually curves spacetime. And it's also part of a unified theory that includes quantum theory as explained in the chapter on Computing Quantum Reality.

1. By the METc Principle mass-energy and space are two aspects of the same thing. They are both forms of spatial velocity in the form of in-place vibrations in the fabric of computational space at a minimum dimensional scale.
2. Empty space is a field of minimal amplitude vibrations corresponding to the zero-point energy of the quantum vacuum, which is likely related to the c value of the speed of light.
3. Gravitational fields are fields of increased amplitude vibrations centered on massive particles.
4. Massive particles are detached concentrated units of space packaged in particle component sets.
5. Packaging in particle component sets enables particles to move relative to the empty space background.
6. The mass of a massive particle is a field of space vibrations centered on the particle component package. The vibrational field is an inseparable part of the mass of particles.
7. Gravitation is homogeneous (same effect in all directions at any point) thus the vibrations must have the same form in all three directions. In a reduced 2-dimension representation they can be visualized as standing symmetrical wave peaks oscillating up and down. In 3-dimensions they are symmetric point pulses of expansion and contraction at an extremely fine scale.

8. The gravitation produced by multiple masses reinforces but never cancels therefore the peaks produced by multiple masses must be evenly distributed in the fabric of space in the same locations and multiple masses additively amplify (increase the height of the peaks) the vibrations which corresponds to the strength of the field.
9. Thus the fabric of space consists of a continuous field of velocity vibrations of various amplitudes corresponding to the presence or absence of masses and their vibrational gravitational fields.
10. The locations of individual vibrational peaks can be taken to define unit cells in the fabric of space. These cells can be considered the individual quanta of space. They represent the minimal computational granularity of space and are far below the scale of particles and quantum interactions.
11. Vibrational space is a flat Cartesian 3-dimensional computational space in which individual point cells can have different velocity densities.
12. Vibrational space is composed of cells of unit nominal dimension. These cells are the basic computational entities of space and are encoded by the cells of the computational array in which dimensional space is represented and computed.
13. Every cell has an individual vibrational *amplitude*, which is its gravitational field strength.
14. A cell's vibrational amplitude gives it a proportional intrinsic spatial velocity, which is the strength of its gravitational field.
15. Every cell has a nominal Cartesian width, which is the same for all cells, and an actual traversal width that is greater in a gravitational field when the ups and downs of its vibrational peaks and valleys are considered. The nominal cell width is the same for all cells but the traversal width varies with the amplitude of its vibrations, the strength of its gravitational field.
16. By the STc Principle each cell has a slower time velocity due to the intrinsic spatial velocity of its vibrations. This is the source of gravitational time dilation.
17. These three effects characterize every cell in computational space and are greater in cells in gravitational fields since they have greater vibrational amplitudes. They account for all general relativistic effects just as the curved space model of general relativity does.
18. The curved space model of general relativity is completely equivalent to our vibrational space model. If the vibrations of a volume of vibrational space were frozen and stretched out that volume would be dilated and the space there curved as a result. Conversely if the space curves of general relativity are

compressed into a landscape of peaks and valleys and set into vibration we get our vibrational model. So essentially our vibrational space model replaces the curved space of general relativity with vibrations in the individual cells of a flat Cartesian space.

19. Both models give exactly the same relativistic effects but our model is superior on three counts. It models space as the flat Cartesian space we actually observe; it provides a unifying explanatory mechanism for gravitation missing from general relativity (why the presence of mass curves space is unexplained in general relativity); and it also models quantum reality as explained further on.

20. Gravitational fields are fields of intrinsic spatial velocity due to their vibrational amplitudes. Thus by the STc Principle an object in a gravitational field experiences the intrinsic velocity of the field and its velocity in time is slowed. This is the source of gravitational time dilation.

21. The total spatial velocity of an object is now its linear velocity plus the intrinsic velocity of its location in a gravitational field. By the STc Principle this total spatial velocity vector subtracted from c reduces its time velocity. Thus the time dilation of linear motion and that of gravitational time dilation are now revealed as two aspects of the same thing, an insight missing from relativity.

22. Objects travel farther through vibrational space than its nominal Cartesian dimensions because the ups and downs of all its vibrations must be traversed.

23. This includes light itself. Light always travels at the speed of light but since it must travel farther through denser vibrational space it appears to be traveling slower than c from the point of view of an observer in empty space. In the extreme of a black hole the apparent speed of light drops to zero for this reason.

24. However the velocity of time of a *local* observer is slowed proportionally to the greater distance traversed thus all observers always measure the local speed of light as c in all cases even though it may appear different to remote observers.

25. Thus all observers experience all their combined spacetime velocity as through time at the speed of light. This is their local frame view. However to get the true picture they must recognize any intrinsic spatial velocity they have and subtract that from their apparent time velocity to get their actual time velocity.

26. This gives the true universal view of all relativistic processes in the observable universe because it's the view of computational space in which they are all actually computed.

27. A gravitational field is a field of vibrational density that falls off by the square of the nominal distance from a gravitating mass. Thus every point in the field has a velocity density gradient with the velocity density greater towards the gravitating mass than away from it. This gradient produces a velocity vector pointing towards the mass that inertial motion tends to follow. This is the actual source of gravitational attraction, which is missing from relativity, which doesn't properly explain why an object at rest in curved space begins to move.
28. By the MEv Principle the mass of a massive particle is modeled as a field of fine spatial vibrations centered on the nominal particle. It's the intrinsic spatial velocity of these vibrational fields in aggregate that slows the time velocity of objects in the field by the STc Principle and so accounts for the effects of general relativity.
29. The vibrations of a particle's mass appear to take the form of a field of standing waves characterized by an amplitude and frequency at every point in its field. The amplitude is the intrinsic spatial velocity of the point, which is the source of its gravitational effect, and its frequency is its clock time velocity, it's internal clock time rate.
30. By the STc Principle the vector sum of a particle mass's amplitude and frequency, its intrinsic velocity in space and its velocity in time, must always equal the speed of light c. Thus particles with different rest masses will have vibrations with different amplitudes. Thus we must assume they will also have slightly different frequencies in accordance with the STc Principle.
31. However the masses of particles are so miniscule any difference in their velocity through time, their frequencies, will likely be undetectable. However this just might show up as anomalies in particle half-lives or other aggregate effects.
32. The rest mass of a particle at rest is fixed. However if it gains linear velocity this adds kinetic energy to the vibrational energy of the particle so its total spatial velocity (linear kinetic energy plus intrinsic spatial velocity of vibrational amplitude) is increased and its time velocity is slowed. The slowing of its time velocity (frequency) increases the particle's intrinsic spatial velocity (vibrational amplitude) so the total vector sum of its vibrational velocity in space and time is conserved. Since the frequency is slowed the amplitude is increased and since amplitude is observational mass the rest mass of a moving particle increases as relativity predicts. Thus the mass of a moving particle increases

because its clock time rate is slowed and this increases its own intrinsic velocity in space, which is its observational mass.
33. The same is true in gravitational fields. A particle experiences the intrinsic spatial velocity of the field, which reduces its internal vibrational clock rate. This in turn increases its own intrinsic spatial velocity in the amplitude of its mass vibrations. This increases the *observational mass* of the particle, which in the case of gravitational fields is the particle's *weight*. This is why objects in gravitational fields have increased weight.
34. Thus the increase in observational mass with linear velocity and the increase in observational mass in a gravitational field are both caused by time dilation and both revealed as the same effect, an increase in observational mass which is what manifests as weight.
35. The frequency of the mass vibrations of particles are their internal clock rates, and their amplitudes are their observational masses. The vector sum of both is computed by the fixed number of processor cycles allocated to computing velocity in space and velocity in time. Additional linear velocity reduces the frequency of the vibrations, which increases their amplitude, which is the observational mass of the particle.
36. Photons are massless and have no internal vibrational structure to compute and thus no internal clock time rates. Therefore all the processor cycles that compute photons go to computing their linear wave motion through space and photons automatically always travel at the speed of light because they have no internal velocity through time. In contrast massive particles have internal vibrational structures that must be computed and so can never travel at the speed of light because they always have some velocity through time.

This outline explains the essentials of how fields of mass vibrations in space produce the effects of general relativity. We can now explain in greater detail how relativistic spacetime is actually computed.

RETHINKING SPACETIME

It will be useful to first take a moment to examine the concept of spacetime before proceeding. The notion of a single fixed universal spacetime within which all things exist and all events occur is absolutely fundamental to science and our common sense view of the world, yet there are very good reasons for thinking it simply doesn't exist, and no evidence that it actually does exist.

For one thing we certainly never observe or measure any such fixed empty space. All we actually observe and measure are the dimensional relationships between interaction events of particulate matter, specifically the events of measurements and observations of objects by us. Try as we may, we simply cannot observe or measure empty space itself. However we try we always end up observing or measuring the dimensional relationships of objects and events 'in' space. Thus our concept of an empty space in which things exist is actually a logico-mathematical construct inferred from the dimensional relationships among the events we measure rather than an actually observable physical structure.

For example we never see the actual empty space between objects and us. What we see is an object's apparent size on our retinas, which is then processed by our brains to compute a distance relationship between us and the object based on a mental model of the presumed and apparent sizes of the object. It's this dimensional relationship that's observed rather than any actual empty space between us and the object. And this is true in all cases without exception.

So the apparently physical space around us is actually our brain's projection of the dimensional relationships generated in our simulation of reality back out into an apparently external world of its own creation. And we know this is a fiction that exists only as data structures within our neurons. The only thing true about it is that these dimensional relationships in aggregate do form a consistent 3-dimensional logico-mathematical structure in which objects can be meaningfully placed by our simulation.

For example if one object is 10 feet away from us and another 30 feet in a straight line then we can correctly compute the objects are 20 feet apart. So spacetime is the consistent logico-mathematical structure that emerges at the aggregate level of dimensional relationships rather than an observable physical structure.

If spacetime is actually a logico-mathematical construct then it needn't exist as the physical entity science assumes, and can just as easily be an information structure that emerges from aggregates of dimensional relationships. So the apparently physical spacetime of science could just be the overall mathematical structure of how dimensional relationships emerge at aggregate scales. There is no way to demonstrate this isn't true.

And a logico-mathematical structure rather than a physical one is all we need for science to keep working as it always did.

General relativity has already gone part way to this understanding. General relativity imagines no single universal space that is valid for all observers. It conceives of space in terms of *manifolds*, which are views of curved space from the perspectives of individual observers (Wikipedia, Manifold). And relativity tells us that there is no single curved background space that all individual manifolds exactly map to. The spacetime manifold of every relativistic observer can be different.

Though relativity conceives of manifolds as *views* of spacetime rather than actual different spacetimes, relativistic manifolds can be inherently inconsistent with each other, which casts considerable doubt on the concept of a single universal spacetime. Certainly a single universal spacetime that is the same for all observers doesn't exist in general relativity.

Of course general relativity can model single cosmological spaces quite effectively, but only when the views of individual observers are artificially ignored and other generalizations are made (Wikipedia, Friedmann–Lemaître–Robertson–Walker metric). So we can very reasonably conclude that what general relativity is really telling us is that there isn't a single universal spacetime valid for all observers. In any case it's clear that a computational spacetime consisting of multiple independent spacetime fragments can be consistent with general relativity.

In view of this evidence it's clear that the single universal spacetime background that science assumes is one more questionable *interpretation* of science, rather than an observable fact of science.

Universal Reality has no difficulty in dealing with the lack of a single universal background spacetime. In Universal Reality spacetime is not a single separate pre-existing container *for* physical events but is the dimensional information we already know is computed *by* events.

Taking spacetime as a logico-mathematical structure that emerges from events rather than a pre-existing container for events solves a number of important conceptual problems and doesn't diminish the applicability of science to explain and predict natural phenomena.

HAPPENING & THE P-TIME PROCESSOR

Universal Reality proposes a computational universe in which a relatively simple elemental program is executed in a manner that automatically generates a universe that obeys the rules of both general relativity and quantum theory. The key to this unity is that both material structures and spacetime are computed together as a single integrated data structure. This model is based on trial implementations of model universes using the XOJO programming system on a Power Mac.

The universe consists solely of the data of particles and their particle components, and the program that computes them. It executes in a computational space identified with the quantum vacuum to update all the data of the universe at every P-time tick, each of which recomputes the complete data state of the observable universe in the current universal present moment.

Happening animates the universe and drives its temporal evolution. Happening operates as a universal processor, which is an innate aspect of the quantum vacuum, the medium or substrate of existence in which all the data that constitutes the universe exists.

The data states of all processes in the universe are computed simultaneously at every P-time tick because all the data of the universe exists simultaneously in the quantum vacuum whose happening is its processor. Dimensional spacetime including the computations of all local clock time rates are computed simultaneously in the current P-time present moment by the processor of happening for all processes in the observable universe.

In the XOJO model this is done in a clock event executed for loop on all particles, but in the actual universe the data of all particles is computed simultaneously because all data exists simultaneously in the processor in computational space. The reasons for identifying computational space with the quantum vacuum and its details have been explained in previous chapters.

In every P-time tick there is a *separate application* of the processor to recompute every distinct *coherent process* in the observable universe. All the applications of the processor compute simultaneously in the current P-time tick so that all processes in the universe are computed simultaneously. Each processor application contains sufficient processor cycles to fully recompute each process in the universe. There are a fixed number of processor cycles allocated between computing velocities in space and velocities in time so that the combined spacetime velocity of all processes is always the speed of light, c.

THE UNIVERSAL REFERENCE BACKGROUND

The computational space of reality is non-dimensional in the same sense that computer programs define non-dimensional computational spaces. The computations themselves are observable only through their effects on the data that makes up the observable universe. All data, dimensional and structural, is computed within the computational space of the quantum vacuum. This data is then interpreted as a physical universe of material structures in a physical spacetime in observers' mental simulations of reality.

Computational space and P-time together are a preferred unobservable background of purely numeric (dimensional data stored are numbers) structure and dimensionality within which all observable dimensional frames are computed consistently. This preferred computational frame neatly solves the problem of what absolute rotation (for example Newton's bucket) is with respect to and what actual world lines are with respect to. They are both with respect to the preferred background frame in which they are actually computed. This is in distinction to the purely observational effects of mutual relative motion, which vanish without lasting effect when the motion stops.

Observers observe the universe in terms of measurements relative to their own frames. All observable measurements ultimately reduce to particle interactions and only particle interactions produce observable data. The actually computed values are with respect to the background frame in which they are computed and are non-observable. Observable values are generated by particle interactions and are always observed with respect to the frame of some observer.

The relativistic effects of motion through space are experimentally confirmed and widely used to correctly calculate the trajectories of bodies in space, but there is a deeper mystery at the heart of relativity that hasn't been solved until now. Namely what is actual spatial motion relative to? There is nothing in relativity theory itself that explains this because relativity claims that all frames are equivalent and none is preferred over any other. This is why it's called relativity and this is considered a basic principle. But this is an incorrect interpretation.

In the twin example why is motion with respect to the earth actual, and produces actual agreed effects, but the motion of earth with respect to the traveling spacecraft isn't and doesn't? The equations of relativity provide no answer, and this has been an unsolved dilemma for over a century since relativity first appeared.

Thus there must be a fundamental assumption in relativity itself that goes largely unrecognized and is even actively denied. There has to be a single absolute fixed background with respect to which actual as opposed to purely observational relativistic effects occur. But the whole original idea underlying relativity was that all motion is relative. So if all motion is relative how can its effects only be actual with respect to some absolute notion of spacetime that relativity can't even properly define? This is an important problem that clearly requires a solution.

This problem arises both in determining the actual spatial lengths of world lines and the actual time dilation of clocks traversing them, and in determining what actual rotation is relative to as explained in the next section.

There is clearly some absolute reference with respect to which actual spatial motion and actual rotational motion are relative to but what is it? Relativity tells us that all coordinate systems are equally valid so why couldn't we pick a coordinate system moving along with the traveling twin's ship and have the clocks back on earth actually slow down rather than the clocks on the spaceship?

There has to be an absolute frame with respect to which actual motion occurs and purely relative motion doesn't. If there wasn't logical contradictions would occur when space travelers meet because they would both see each other's clocks still ticking at different rates standing right next to each other.

This would lead to all sorts of problems with the laws of physics. Which clock rate would actually describe which laws of physics at that location? The twins would both age at different rates in front of each other's eyes and physical processes could run at wildly different rates in the same location with disastrous results. Thus some sort of absolute spacetime is required to maintain the logical consistency of the computations of the laws of nature and keep a computational universe from tearing itself apart.

As Ernst Mach pointed out in the case of rotation it appears this absolute frame is more or less aligned with respect to the total mass of the universe and this is roughly in accord with observational results. But why? This is a fundamental question few physicists have attempted to understand so it's usually just ignored if even recognized.

More recently the cosmic microwave background (CMB) radiation is being used a reference frame for cosmological motion but the fundamental problem is still the same. Preferred frames have crept back into modern physics over the protests of strict relativists, even though the very concept is antithetical to the originating concept of relativity.

It's also difficult to understand how a uniform effect throughout the universe could be the result of the obviously non-uniform distribution of mass in the universe at galactic scales. How could the total average mass of the entire universe cause the effect and not the much nearer distribution of mass in our galaxy or even our solar system? It doesn't seem to make sense.

However the reason for Newton's bucket and the privileged alignment of spacetime with the aggregate mass of the universe is a natural consequence of how spacetime emerges computationally in Universal Reality.

The reason there is a privileged background for both effects is because spacetime is created by quantum events as explained in the chapter on Computing Quantum Reality. At the largest scale the interactions of all particles in the universe form a universal network of entangled dimensional relationships with respect to which the dimensional relationships of all subsequent events naturally align as they are computed with respect to it.

This network of dimensional relationships is a consistent logico-mathematical framework with respect to which the dimensional relationships of subsequent particle events automatically align. This is the solution to Newton's bucket and the reason why world lines with respect to the aggregate mass of the universe is approximately the correct choice that produces actual as opposed to observational relativistic effects.

Relativistic events take place with respect to a more or less absolute spacetime background because they are computed with respect to the dimensional entanglement network. So instead of being an empty *physical* structure, spacetime is instead a *logico-mathematical* structure in which subsequent computations automatically align with the ones from which they are computed. The mass distribution of the universe is an aspect of the entanglement network so subsequent dimensional computations occur with respect to it.

Thus the dimensionality of the entanglement network becomes an absolute reference background with respect to which actual as opposed to merely observational relativistic effects are relative. Without this absolute reference background there could be no notion of actual as opposed to observational relative linear or rotational motion and relativity itself would not be consistent. Effects such as the twin paradox and Newton's Bucket would lead to contradictions and the observable universe would likely cease to exist.

Since the absolute reference background is continually recomputed at the local level there could be small inconsistencies from location to location and from era to era across the universe, which might produce slight anomalies in the rotations of gyroscopes or the expected time dilation of long space flights or other relativistic effects. See the upcoming section on Dimensional Drift for more on this.

So the correct definition of absolute rotation or linear motion is with respect to the dimensional consistency of the *proximate* absolute reference background, the dimensional alignment of the entanglement network. In general this will be very closely through not necessarily exactly aligned with the total mass distribution of the universe.

The absolute reference background may have other implications. Given their intimate connection with spacetime could the values of c and G, the speed of light and the gravitational constant, be somehow a function of the total mass of the universe including that beyond the particle horizon and could the size of the observable universe then be determined from those values? Could c and G both be emergent effects of the dimensional entanglement network rather than intrinsic constants of the complete fine-tuning?

NEWTON'S BUCKET & MACH'S PRINCIPLE

This universal reference background is the key to understanding two fundamental problems of relativity. It provides the solution to the problem of Newton's Bucket and the question of what relativistic world lines are relative to.

When a bucket of water is rotated the water begins to rotate and climb the walls due to centrifugal force. This is easy to understand but there is a hidden mystery involved. What is the water rotating with respect to? A bucket of water that is not rotating has a flat surface, but what is the water still with respect to in that case (Wikipedia, Bucket argument)?

At first we might suspect the rotation is with respect to the surface of the earth but that isn't true because we have the same effect in the rotation of gyroscopes in deep space. They are either rotating or they aren't, and they always have a rotation that must be relative to something but there is no known physical mechanism in science to explain what.

It's clear there is some absolute reference spacetime that more or less aligns with the total mass of the universe, and the rotation of the bucket is with respect to that, but there is no law of physics that specifies why actual rotation is always with respect to the CMB or total mass of the universe rather than any other coordinate system.

Ernst Mach proposed that the rotation was with respect to the distribution of inertial mass of the universe and was some as yet unknown effect of gravitation but he could offer no scientific reason for this and there isn't any law of science that provides any reason why it should be true (Wikipedia, Mach's principle). And just making up such an important fundamental law that has no connection with any other law and has no other apparent effects is clearly unjustified.

Once again the explanation comes naturally to Universal Reality from its concept of an absolute reference background in which the universe is actually computed. It seems quite obvious that actual rotation must be relative to the established consistency of the frame in which it's computed, as opposed to some observer frame. If for example an observer is riding a rotating carousel in empty space dark the carousel appears to be motionless but the presence of centrifugal force proves it isn't.

So the solution to Newton's bucket is simple. Actual as opposed to observational relative motion is always with respect to the dimensional consistency of the universal computational space in which all dimensionality and motion is computed.

The quantum of rotation is spin. Spins are the elemental units of rotation and angular momentum. Thus as dimensionality is computed the resulting network of entanglement relationships gains an absolute dimensional *orientation* due to the progressive alignment of the spin orientations of all non-zero spin particles with respect to the computational background. As a result the dimensional network of the observable universe becomes a common absolute orientation reference as it's computed. All rotation is relative to it because all successive computations are computed from previous ones in terms of them.

This doesn't mean the axes of spin orientations are all pointing in the same direction, but that they are all pointing in *some* direction relative to a common reference standard of orientation across the entanglement network. If there was no absolute reference standard for orientation rotations couldn't even be compared.

So the existence of spin is another little necessary ingredient of reality because in aggregate it generates a universal absolute reference orientation standard with respect to which all rotation is relative. So it may be only the existence of the spin particle component that in aggregate builds a universe in which rotation makes consistent sense.

Thus Universal Reality's computational approach to spacetime provides a solution to the problems of Newton's bucket, what world lines are relative to, and why there is an absolute underlying frame in spacetime that explains actual versus observational relativistic effects. And it does this in a manner that leads to a unification of relativity and quantum as explained below.

This same approach also resolves the apparently paradoxical nature of quantum processes, as we will see in the upcoming chapter on Computing Quantum Reality. The universal reference background that emerges computationally from quantum events provides a dimensional model that explains general relativity but is fuzzy enough to accommodate quantum effects as well.

DIMENSIONAL DRIFT

Dimensional drift is the hypothesis that the absolute background relative to which actual relativistic effects occur may not be completely consistent from location to location because it's computed locally by quantum processes that are inherently dimensionally fuzzy. It also clearly changes over time with the redistribution of particles in the universe due especially to the expansion of space and this could also be a source of anomalies in its consistency from one time and place to another.

Thus at very large scales the exact dimensional parameters of the universal background frame may not be exactly cross-consistent or correctly known from earth. Attributes such as scale, reference motionlessness, and orientation of the absolute background could theoretically vary from one location in space to another. So it's possible we might observe unexpected anomalies in relativistic measurements from one location to another that could provide a conclusive and falsifiable test of Universal Reality versus general relativity.

However such anomalies might be difficult to detect because our dimensional measurements could be subject to the same anomalies depending on their nature. However if one location of the absolute background was either stretched or moving slightly with respect to another and a space probe went from one location to the other its signals might indicate unexplained relativistic effects.

Such small anomalies have in fact been actually detected. For example the two Pioneer spacecraft seem to be slowing slightly more than relativity predicts as they leave the solar system. While the currently accepted explanation of the slowing is the thermal recoil force from onboard generators this doesn't seem to explain small variations in the effect and there have been a number of other explanations proposed (Wikipedia, Pioneer anomaly).

Since the dimensionality of the observable universe is computed locally at the particle scale it's certainly reasonable to assume the alignment of the absolute background with the average mass distribution of the universe is only approximate and may vary slightly from location to location. Using the total average mass of the universe should provide the same absolute reference frame everywhere in the universe, but the absolute background could vary slightly from area to area. What should be definitive is the logico-mathematical consistency of the entanglement network whose dimensionality is the absolute background at the local scale. This is the reference with respect to which spatial motion actually occurs, and that may or may not be in exact alignment with the total mass of the universe or the CMB.

The absolute reference background is not a fixed pre-existing spacetime container, nor even a fixed data structure, but is the aggregate logico-mathematical consistency of all individual dimensional events as they are continually recomputed over time. Thus the consistency persists but the dimensionality that supports it clearly evolves over time as the motion of cosmic masses are computed and the universe expands. The current Hubble expansion is very slow, however in the inflationary period the expansion seems to have been enormous and near instantaneous and its dimensional effects might persist long after.

The first particle events began occurring with the big bang and inflation and since these earliest events began consistently the logico-mathematical background of their aggregate consistency immediately emerged as an implicit information structure relative to which the dimensionality of all subsequent events could be said to occur.

Thus the standard reference background of computational spacetime with respect to which absolute linear motion and rotation are relative in the sense of producing actual relativistic effects is constructed piecewise by individual quantum events whose dimensionality is inherently fuzzy.

So relativistic dimensionality should have manifested from the beginning or at least almost from the beginning without any problems, however the overall reference consistency it's relative to has clearly changed over time with the dimensional evolution of the universe.

The absolute reference background reflects the aggregate dimensional consistency among all events in the observable universe, but the distribution of events continually changes over time and varies across the universe. On average the dimensional background remains the same but it's possible that some local differences in the background might sometimes develop. The overall homogeneity of the dimensional background could be subject to local distortions and dimensional drift.

As space expands and the distribution of galaxies changes there could be some measurable drift of the background dimensionality with respect to which absolute motion along world lines occurs. This might be detectable as anomalous astronomical effects in particular with respect to absolute linear and rotational motion.

This is something Universal Reality predicts could occur, and if it does would provide good evidence for Universal Reality's computational theory of spacetime and an absolute background reference dimensionality with respect to which actual motion occurs.

If in fact there are areas with sparse enough or different enough computational connections that have slightly different dimensional background references there is surely a computational process that reconciles them when they intersect. Only in this manner could a consistent computational universe be maintained. But of course the quantum vacuum already has a mechanism to create and choose among probability distributions via the process of random choice so it's to be expected it must have some general computational self-correcting mechanisms to enforce its overall logico-mathematical consistency.

Most likely any such anomalies on the cosmic scale are currently misinterpreted as either slight inconsistencies of measurement or misleading dimensional measurements. But if interconnected effects were to be compared and inconsistencies discovered this type of error might be detectable.

Another possible source of dimensional drift may be due to the granularity of dimensionality in a digital universe. The universe should be granular and digital at its minimum scale because only exact digital data can be consistently computed. But this precludes infinite precision in dimensional values and their calculations. Any inconsistencies here are presumably reconciled by taking numeric averages at the finest scales. However over time rounding errors could conceivably accumulate to dimensional discrepancies at observable scales even though the scale of granularity is likely many orders of magnitude below that of even quantum scale measurability.

It is not altogether clear the ways dimensional drift might show up. A uniform expansion or contraction of the background wouldn't be observable but we see the past as well as the present so if it changed significantly over time observable discrepancies in relativistic measurements might be detectable between different eras in time.

It is clear from relativity itself there is a preferred background frame with respect to which actual relativistic effects occur. This frame is established by the logico-mathematical consistency of all computations of the entanglement network, and thus closely aligned with the distribution of matter in the universe as Planck surmised. However all individual events are local so absolute motion should be with respect to local areas of the overall logico-mathematical consistency. Ultimately any discrepancies of consistency should depend on variations in the flux density of particle interactions over time across the universe.

The big question is to what extent this background reference and its actual relativistic effects is local to the mass-energy distribution of our galaxy or perhaps even to some extent our solar system rather than to the average distribution of all mass-energy in the universe. This likely depends on the density and connectedness of events across the universe relative to what degree mass-energy distributions have changed over time. This should lead to testable predictions.

Thus if an observer is stationary with respect to at least the local consistency of the logico-mathematical background as human observers are for most practical purposes the relativistic effects they observe are actual with slight corrections for their own relative motion with respect to it.

An absolute background and dimensional drift are clearly testable proposals. If we find the absolute background reference dimensionality against which actual world lines are measured is slightly inconsistent across different regions this would be strong evidence for Universal Reality's model of reality.

UNIFYING RELATIVITY & QUANTUM THEORY

A major goal of Universal Reality is to outline a conceptual model that unifies quantum theory and general relativity. The fundamental inconsistency between quantum theory and general relativity derives from their different concepts of spacetime. Quantum theory considers spacetime a fixed pre-existing container for events that remains unmodified by events. However the spacetime of general relativity is dynamic and is curved by the presence and movements of mass-energy even though it's also a pre-existing container for events.

Thus it's clear that to unify the two theories their divergent views of spacetime must be reconciled. The key to achieving this is to recognize how quantum events actually create dimensionality so that the dynamic spacetime of general relativity naturally emerges from them. In this model spacetime is the dimensional relationships among events, and is computed simultaneously with the events.

This approach is suggested by several lines of reasoning, two of particular importance. First when we carefully consider the nature of spacetime we find it reduces to the dimensional consistency among measurements. In other words we never actually observe empty space but only the dimensional relationships among mass-energy structures. All we actually observe is particle-based events with dimensional values. This includes the measurements our senses make as we experience the world. Thus what we think of as spacetime is actually the logico-mathematical consistency among dimensional measurements and there is no observational evidence that empty space as an encompassing physical container for events even exists.

Second quantum events conserve mass-energy and due to the non-proportionality of particle masses energy excess energies must be expressed in terms of velocities of emitted particles. Thus the non-proportionality of particle masses requires the creation of dimensionality for events to occur. Particle events must create dimensional spacetime to be able to occur so total mass-energy can be conserved. The conservation of mass-energy in quantum events must be expressed in terms of velocities and their resulting positions, which are defining elements of spacetime. Thus by creating interrelated positions and velocities vast chains of particle events create the network of dimensionality observers interpret as spacetime.

Taking these points together in the context of a computational universe in which everything is ultimately numeric data, suggests that quantum events are actually creating spacetime on the fly as a mechanism to conserve mass-energy.

And only if spacetime is dynamically created by events rather than being a pre-existing container for events, can the dynamic spacetime of general relativity be generated by quantum events. So this is a very promising approach.

This model enables the wavefunctions, uncertainty, and vacuum energy fluctuations of quantum theory to be reinterpreted as descriptions of spacetime in the process of creation rather than descriptions of fuzzy particles with respect to a fixed pre-existing infinitely exact spacetime. The quantum equations, in so far as they are accurate, remain the same but their interpretation is turned on its head.

In this model quantum fuzziness or indeterminacy isn't the innate nature of particles but the innate nature of spacetime as its being computed. This in turn implies that particles and particle events could be fairly simple arithmetic data, and the complexities of quantum theory emerge only as dimensionality is computed. Note that this view is consistent with the underlying trajectories and time evolution of wavefunctions, which are exact. This implies a more fundamental exact data description on which dimensional fuzziness is overlaid.

So if reality might be numerically exact at the fundamental level, why is dimensionality fuzzy at the quantum level? What is the mechanism that produces quantum indeterminacy when dimensionality is computed?

It turns out the manner in which the processor computes the data of the universe provides a reasonable mechanism for both the STc Principle that underlies relativity and quantum indeterminacy. The allocation of a fixed number of processor cycles between calculations of spatial velocity and time velocity is the source of the STc Principle, and fine scale processor cycle oscillations between space and time can account for quantum fuzziness.

PROCESSOR CYCLES & THE STc PRINCIPLE

In every P-time tick the processor of happening runs the elemental program against each coherent data state in the universe to compute a process. Each process creates an event in each P-time tick. Particle interaction events produce potentially observable values but most events are virtual and merely compute the unobserved evolution of particle data. The elemental program analyzes each data state to determine which of its subroutines is required to compute the current process and branches to that subroutine.

A separate *application* of the elemental program computes each separate coherent process in every P-time tick. Each separate process is computed by many processor cycles in each P-time tick to produce its next current present moment data state. A fixed number of processor cycles (perhaps more accurately sets of cycles but for convenience referred to herein as cycles) in every P-time tick are allocated to compute the total spacetime velocity of every process. This fixed number of cycles is allocated between computing the velocity in space and velocity in time of each individual process. The fixed number of velocity allocated processor cycles in each P-time tick sets the value of c, the speed of light.

These velocity allocated processor cycles are the computational source of the STc Principle; that everything in the universe continually moves though combined space and time at the speed of light. More precisely the vector sum of the space and time velocities of everything in the universe is always c as required by relativity. Thus the fixed cycle rate of the universal processor is the source of the value of c and its allocation between space and time velocities the source of the STc Principle that underlies the relativistic nature of reality.

The total spatial velocity of a process, its linear velocity plus the intrinsic velocity of any fields, is computed first, and the cycles left over then compute the internal evolution of the process. The rate of internal evolution of a process manifests as how fast the process is happening, which is observationally its proper clock time rate. Fewer processor cycles left over to compute the evolution of internal processes produces a slower clock time rate for the process. This is the computational source of clock time and its relativistic time dilation.

The processor of happening computes all local clock time rates process by process simultaneously in each universal P-time tick. Thus all the different local clock time rates in the universe are computed simultaneously in the current universal present moment common to the entire universe.

The processor cycle rate has no explicit rate itself because it's the source of all relativistic clock time rates and can only be measured in terms of the clock time rates it produces. The processor computes all processes in the universe to have a fixed speed of light velocity. This total velocity is distributed between spatial velocity (linear velocity plus intrinsic velocity of any fields) and the internal rate at which a process evolves which is its clock time rate, its velocity in time.

PROCESSOR CYCLES & QUANTUM RANDOMNESS

This computational model in which the universal processor allocates a fixed number of cycles to computing velocity in space and velocity in time also explains wavefunctions and other aspects of quantum indeterminacy in a straightforward manner.

Assume the allocation of processor cycles between space and time computations at the quantum level is not precise but subject to very fine random oscillations. Effectively this means the dimensional background in terms of which dimensional values are computed is not fixed but continually jittering. The background rulers and clocks used to compute dimensional values are continually oscillating and exchanging identities during each P-time tick.

As a result dimensional values involving space and time related variables (for example energy and time, or position and momentum) are computed in terms of a fluctuating standard and are always mutually uncertain by a minimum amount described by the Uncertainty Principle. So this intrinsic indeterminacy of space versus time at the quantum scale as dimensional values are computed is the source of the Uncertainty Principle.

Modern physics derives the Uncertainty Principle from the zero-point energy fluctuations of the quantum vacuum and this is consistent with our processor cycle model (Wikipedia, Uncertainty principle). The quantum vacuum is a direct manifestation of the virtual reference background of spatial dimensionality. The zero-point energy fluctuations are a direct manifestation of the processor cycle oscillations between space and time.

As a result each point of the quantum vacuum appears to continually oscillate between energy (velocity in space) and velocities in time. Energy excitations (velocities) in space continually appear as virtual particles but on time scales (velocities) so short they vanish in most cases before they can fully actualize. As a result the quantum vacuum is a fluctuating sea of transient virtual particles with an average energy determined by the spatial aspect (velocities) of the processor cycle oscillations.

These same processor oscillations can also explain why quantum particles appear to behave as wavefunctions. Whenever the exactly conserved energies of particles is translated into dimensional values of positions and velocities the processor cycle oscillations give them indeterminate values in a manner that evolves according to the Schrödinger equation (Wikipedia, Schrödinger equation).

The processor cycle oscillations for each separate coherent process are independently computed so that the processor imposes oscillations of different phases on each process scaled by the intrinsic velocities of each process, such as the internal vibrational velocities of mass, and the nominal trajectories and velocities of the underlying numeric particle modeled by the wavefunction. As a result each coherent process develops a scaled indeterminacy around its nominal trajectory that evolves along its nominal trajectory and this indeterminacy takes the mathematical form of a wavefunction. The overall velocity and trajectory of the wavefunction is calculated by the apportionment of processor cycles, while the indeterminate waveform of the wavefunction is produced by the processor cycle space versus time oscillations.

Thus the universe at the quantum scale is accurately described by wavefunctions that are mistakenly thought to encode the dimensional indeterminacy of particles with respect to a fixed exact spacetime background. Modern physics interprets wavefunctions as particles fuzzy with respect to a fixed spacetime background. However Universal Reality interprets wavefunctions as fragments of incipient dimensionality inherently fuzzy with respect to each other due to the manner in which processor cycles compute dimensional values.

This new interpretation of wavefunctions as fluctuations of spacetime itself as it's computed rather than particles smeared out in a fixed preexisting spacetime frees us from the model of spacetime as a fixed container for events. This in turn enables spacetime to be seen as emerging from the dimensionality of mass-energy structures and computed with them as a single unified structure. What we interpret as a dimensional spacetime now consists of dimensional relationships among particles that are computed by quantum events. This enables a single unified process to compute a universe that obeys the laws of both quantum theory and general relativity, and models mass-energy structures and their dimensional relationships as a unity as well.

Like relativity all types of quantum indeterminacy are a natural result of the allocation of processor cycles in our computational model. As we will soon see both relativity and quantum theory now emerge automatically and consistently from this single computational model, general relativity at large scales, and quantum phenomena at fine scales. This model also explains the underlying phenomena of quantum theory in a completely non-paradoxical manner.

FOUR COMPUTATIONAL LEVELS

Our proposed model of the universe can be better understood in terms of four computational levels. Each level is a perspective on the unified mass-energy spacetime universe as it's computed. This four level view is conceptually useful though everything actually exists as a single unified process.

0. At Level 0 the universe consists entirely of the exact numeric data of all particle components and particles being computed by applications of an elemental program that consists of fairly simple logical and arithmetic operations. These computations take place in the computational space and universal P-time by the universal processor of the quantum vacuum in accordance with the virtual data of the complete fine-tuning. This level is not directly observable.

Every coherent set of particle interactions, including particle-field interactions, is computed as an individual process. There is no physical universe and no actual physical motion or dimensionality or mass-energy structures at this level. Everything exists as numeric and relational data states whether dimensional or non-dimensional. This actual fundamental level of the universe is computational and completely non-physical and non-material.

All the data that constitutes the universe exists simultaneously in the quantum vacuum, which acts as a universal processor that continually computes its current state from its previous state. Each universal computation manifests a current universal present moment P-time common to all processes. This universal computational space and P-time establishes an absolute reference background for all the dimensionality and mass-energy structures of the subsequent levels.

At every P-time tick a separate application of the single fundamental program uses multiple processor cycles to recompute each coherent process in the universe. Each P-time tick manifests as the current universal present moment of the universe in which all clock times, mass-energy states, and their dimensional relationships are computed individually on a process-by-process basis.

An important consequence of this model is that nature doesn't have to store data as equations as it would have to do if wavefunctions and dimensional fragments were actual data components of the universe. Everything is now stored and computed as its exact Level 0 particle component data values. Thus the observable universe can consist only of exact data values without any equations necessary. This greatly simplifies a computational universe and it's reasonable to assume nature strives for simplicity.

1. Level 1 is the quantum universe that physics describes with wavefunctions, the Schrödinger equation, and other quantum equations. The quantum wavefunction picture of reality is due to

the processor allocating the cycles it uses to compute velocity in space and velocity in time in a manner that oscillates between space and time at the quantum scale. This results in a spacetime indeterminacy of quantum processes that manifests as wave-like probability distributions of space versus time of complementary dimensional variables such as position and velocity (momentum).

As a result whenever any dimensional value is computed it's inherently uncertain with respect to its space and time dimensionality at quantum scales. In effect the space versus time rulers with respect to which events are measured are continually oscillating back and forth between space and time as the dimensionality of events is created.

These processor cycle oscillations have a number of important effects with respect to space and time dimensionality. They are responsible for the apparent wave nature of particles, and for the Uncertainty Principle and the zero-point energy fluctuations of the quantum vacuum from which it derives. The processor cycles themselves are also the source of the vibrational velocity that characterizes all forms of mass and energy. In other words they are an analogue of the Higgs field as explained in the next section on the equivalence of mass-energy and space.

2. Level 2 is the observable universe. It consists of the totality of all measurements of observables. Measurement of observables includes the decoherence values of all particle interactions which are effectively mutual measurements by particles of each other's dimensionality. All scientific measurements and observations ultimately reduce to these individual particle decoherences. Level 2 also includes all types of sensory observations, which also ultimately reduce to particle decoherences.

Note that while wavefunctions and the wave nature of particles are useful mathematical structures they are not directly observable and thus not part of the observable universe. For example the observables in the double slit experiment are individual point particle impacts on the back screen (Wikipedia, Double-slit experiment). The waves that produce them are never directly *observed* but only *inferred* from the pattern of the point particle impacts. Thus the wave nature of particles is inferred from patterns of particles behaving as particles and is not an actually observable phenomenon.

The observable universe consists of an entanglement network of observable values produced by the total interconnected network of particle events. The entanglement network is produced by the conservation of particle components through all particle events. This produces consistent relationships among the particle component values of particles emitted by events, and by extension of all particles in the universe since all observable particle values are backwards connected by events through the entanglement network into the single consistent structure of the observable universe.

The entanglement network is the combined spacetime mass-energy data structure of the observable universe. It incorporates the combined data of all mass-energy structures and their dimensionality in a single consistent logico-mathematical structure. By a consistent logico-mathematical structure we mean that the relationships among all observables obey the consistent logico-mathematical laws of physics insofar as they are known.

General relativity is incorporated into the quantum entanglement network with the addition of two elements. First the dimensionality that emerges from quantum measurements is scaled by the velocity equivalents of mass-energy. The equivalence of spatial velocity and mass-energy has been explained above.

Second the allocation of a fixed number of processor cycles per P-time tick between computing velocity in space and velocity in time results in time dilation (slower clock time velocity) with increased spatial velocity in accordance with general relativity even for quantum processes. When all forms of mass-energy are taken as various types of spatial velocity proper clock time rates are computed locally for each process simultaneously and the curved spacetime (velocity density in our model) of general relativity is automatically produced.

So the incorporation of general relativity into the entanglement network of quantum measurements derives from two principles. First treating all forms of mass-energy as various forms of spatial velocity, and second the allocation of a fixed number of processor cycles to computing first velocity in space and then allocating the left over cycles to computing velocity in time. When gravitational mass fields are modeled as fields of vibrational velocity centered on particles this correctly produces the dynamic spacetime of general relativity.

Thus in this model dimensional spacetime is computed together with the dimensionality of mass-energy structures as an integrated system. By abandoning the notion of spacetime as a preexisting empty container for events and letting quantum events compute it we arrive at a spacetime compatible with both quantum theory and general relativity.

So a simple computational process with a processor that allocates a fixed number of cycles between computing space and time is sufficient to compute the entire observable universe as all complexity emerges automatically from the complete fine-tuning as the fundamental program computes the interactions of elementary particles and their particle components.

Level 2 is still not a physical universe. It's the observable universe of scientific measurement that consists only of the consistent universal network of all values of individual observables. It does however consistently include all quantum observables in an entanglement network that implicitly incorporates the curved spacetime of general relativity.

3. The 'physical' universe. The apparently physical universe in which we seem to experience our existence is not actually physical or material at all. Its apparent physicality is our mind's interpretation and generalization of the logico-mathematical consistency of the observable universe of Level 2. Spacetime is a basic aspect of the simulation of reality produced by the human brain and projected outward around us. It's a reification of the data structure of the observable universe of Level 2. Basically the dimensional values of all observables are laid out in the simulation and sorted by their values and an imaginary empty spacetime is interpolated around them to connect them and mistaken for an actual physical entity.

We experience our simulation of the universe as a physical spacetime filled with material objects but this is a convenient illusion. It's a highly successful evolutionary adaptation that greatly simplifies our mental computations of reality that is shared among humans and in various forms by other species. This is explained in detail in the chapter on The Simulation in my book Universal Reality (Owen, 2016).

THE EQUIVALENCE OF MASS-ENERGY & SPACE

Universal Reality models all forms of mass-energy as different types of relative motion (vibrational, waveform, or linear) whose values are their spatial velocities. Only if all types of mass-energy are forms of the same thing can they be converted into each other and conserved. Energy is conserved in all cases as equivalent amounts of spatial velocity are converted from one form to another.

Further space and all types of mass-energy are forms of the same underlying entity, velocity (near null velocity in the case of empty flat space). Kinetic energy is linear velocity, mass is vibrational velocity, and photon energies are wave frequency velocities. Empty space consists of the fluctuation velocity of the quantum vacuum, which effectively opens space up and gives it expanse. The various charges of particles including mass are equivalent to localized vibrational fields in space, and space is the underlying substrate or raw stuff of particle masses. Quantum field theory (QFT) also treats particles as local field excitations (Wikipedia, Quantum field theory).

Particle mass is modeled as a fine vibrational velocity that adds intrinsic spatial velocity to its surrounding dimensionality in the form of a velocity field centered on the particle. Thus gravitational fields are areas of increased spatial velocity surrounding masses that fall off by the square of the radius due to the 3-dimensional geometry of space. Particle charges including mass are not points but diffuse spatial fields of vibrational velocity centered on the nominal location of their point particles. Space itself is the underlying stuff of which charges including mass are composed. Space itself is a single universal field of relative motion, of spatial velocity. Energetic charges are just fields of the same stuff as space with increased localized spatial velocity.

Thus space and spatial velocity are aspects of the same thing. Velocities are energy excitations in space and empty space is the zero-point energy of the quantum vacuum. Particles are localized excitations of the quantum vacuum, of space, that have crystallized around particle components and become able to move relative to the background fabric. This is demonstrated by the Unruh effect in which the acceleration of an observer in empty space manifests as the appearance of particles in surrounding space (Wikipedia, Unruh effect).

While the Unruh effect is an observational effect due to the relative velocity of an observer, it does confirm the equivalence of space and mass-energy velocity that can manifest as particles when instantiated in valid particle component sets. Actual particles are combinations of relative velocity of various forms in space corresponding to the type of charges they carry. Charge fields are fields of various forms of vibrational velocity in surrounding space that add intrinsic spatial velocity to points in the field which reduces the velocity of time in accordance with the STc Principle. Thus other particles passing through these velocity fields experience the time dilation effects of general relativity.

The concentration of sufficient energy (velocity) at points in the quantum vacuum (space) causes actual particles to appear. A particle-antiparticle pair is created if the velocity energy is equal or greater than the mass of the particle pair created so there is sufficient energy velocity to be converted into mass velocity. The concentrated velocity energy is converted into the mass of the new particles with any left over converted into the kinetic energy of linear velocity of the particles. When particle-antiparticle pairs appear all other particle components are conserved since antiparticles have opposite particle component values to regular particles so they all sum to zero. Thus the quantum vacuum is a reservoir of the particle components necessary to make particles when sufficient spatial velocity (energy) is added.

This is an example of the equivalence of mass-energy and space that demonstrates they are aspects of the same thing. All mass-energy is just space with relative velocity of one type or another, and empty space is the diffuse medium of virtual particles from which actual individual particles can be formed.

Fields are aspects of particle charges (including mass as the gravitational charge), rather than separate entities. The spatial velocities of charges are not points but fields centered on the location of the nominal point charge. Due to the fact that particles are energetic excitations in space (velocities) the fields of the four forces can equivalently be modeled as particle exchanges. For example modern particle physics models electromagnetic fields as exchanges of virtual photons (Wikipedia, Quantum electrodynamics).

The equivalence of mass-energy and space when all forms of mass-energy are treated as different forms of spatial velocity takes us a considerable distance towards a new interpretation and better understanding of general relativity.

If particles are not actually points but velocity fields in space centered on points this eliminates the problem of how mass curves space. There is no more mysterious 'action at a distance' in relativity in which mass somehow curves space distant from it. Instead every massive particle extends out through space to the limits of its effects in the form of its intrinsic velocity density field. Particles are intrinsic velocity density fields in space centered on localized points of particle component sets.

A NEW MODEL OF GRAVITATION

In Universal Reality all relativistic effects derive from the equivalence of mass-energy and spatial velocity, and the fact that the vector sum of space and time velocities is always c. This enables us to model combined relativistic mass-energy and spacetime in an entirely new and easy to understand manner. This model is much superior to the usual curved space model of general relativity because it reflects spacetime as Euclidean as we actually observe it and seamlessly incorporates spacetime and mass-energy in a unified structure wholly equivalent to the curved space model of general relativity.

In Universal Reality masses, like all charges, are very fine localized vibrations or excitations of space itself. Thus masses are vibrational spatial velocities and that's all they are. The gravitational fields of masses are very fine vibrations in space surrounding the location of their particle component sets. Charges and fields are not separate entities but fields are inseparable parts of charges. Thus a mass is a field of very fine vibrations in space centered on the point where the nominal particle mass is located.

Note that representing charges as ultra fine spatial vibrations is reasonably consistent with String Theory's representation of elementary particles as vibrating strings though in Universal Reality it's charges rather than particles that are vibrations, and they are different vibrational forms in 3-dimensional space rather than extra compacted dimensions (Wikipedia, String theory).

Because the field itself consists of vibrations in the fabric of space the field increases the intrinsic velocity density of space within it. It adds intrinsic spatial velocity to all points within the field and by the STc Principle this automatically slows time so that total spacetime velocity remains equal to the speed of light, c.

This is a simple and easy to understand model of how gravitational fields produce time dilation. They slow time because they actually are fields of intrinsic spatial velocity. Thus the total time dilation and other relativistic effects of any process is now just a matter of adding its linear velocity and gravitational intrinsic vibrational velocity. Linear velocities and gravitational fields produce the exact same relativistic effects because they are both forms of spatial velocity subject to the STc Principle.

This is an entirely new and revolutionary understanding of general relativity derived from Universal Reality's METc Principle. By recognizing mass as vibrational velocities in space this principle reveals that all relativistic effects derive from the fundamental fact that the total velocity of all processes is always c, and that space itself, in particular its mass excitations are forms of intrinsic velocity that simply add to the total spatial velocity of any process to reduce its time velocity.

This suggests a much simpler conceptual model of spacetime than the curved spacetime of general relativity. We can now model computational spacetime as Euclidean (flat) with each point being characterized by a spatial velocity density. In empty space this is the velocity due to the zero-point energy and the presence of any gravitational or other energetic fields just adds to this velocity density.

Every point in this flat Euclidean space has an intrinsic spatial velocity. This indicates the proportion of processor cycles allocated to compute the time versus space velocity of a particle at that point. So adding this intrinsic velocity to the linear velocity of a particle at this point gives total spatial velocity, which determines the resulting time velocity, the time dilation, of the particle. All relativistic effects become simply a function of the sum of linear and intrinsic spatial velocity of an object at any point.

So we arrive at a very simple and easily understood but completely accurate model of general relativity's spacetime. It looks exactly like ordinary flat space but a gravitational velocity density characterizes each point. The velocity density can be visualized by a velocity meter which indicates the tilt angle of total spacetime velocity from all time velocity at 90° straight up to all in velocity in space at 0° horizontal. The relativistic gravitational effects on particles traveling through points in space depend on the tilt of the meter.

Thus the value of zero-point energy sets the reference point for the tilt in empty space to 90° which corresponds to c, the maximum possible velocity though time. The zero-point energy can be thought of as the resistance to velocity of space or more accurately the maximum possible velocity through spacetime. It's closely related to the speed of light c value of the universe because it's related to how fast space and time can be traversed.

This flat Euclidean model is topologically equivalent to the curved spacetime of general relativity because either can be distorted into the other without tearing but it's much easier to visualize and understand. The curves of curved space can be compressed into vibrational peaks, and the vibrational peaks of velocity density space can be smoothed and stretched out into the curves of general relativity's space.

Velocity dense space looks exactly the same as flat space but particles traveling through it have to move farther because they have to ride the ups and downs of the vibrational ripples just as they have to travel further around the curves of spacetime in the general relativity model. As a result the apparent speed of light through both velocity density and curved space appears slower from the outside though the speed of light always remains the same when measured locally.

One great advantage of the velocity density model is that it reflects the way we actually see gravitational space in the real world as Euclidean. Even if general relativity takes space as curved it still appears Euclidean because light beam trajectories ride the curvature of space. And velocity density space is also amenable to being represented and computed simply in terms of a standard data array in Computational Space.

All in all our velocity density model is much easier to visualize and understand than the equivalent curved spacetime model of general relativity. In fact our model is the most accurate model of the seemingly flat spacetime we observe around us which also appears Euclidean. It's the curved spacetime model of general relativity that while useful and accurate, is misleading and essentially impossible to visualize.

GRAVITATIONAL ATTRACTION

It's clear how the intrinsic velocity of a velocity density field surrounding a massive object causes clocks to slow in accordance with the STc Principle, but how does this explain gravitational attraction? Why do other masses tend to move towards areas of greater velocity density?

The same question arises in general relativity's curved space model of gravitation. Relativity states that inertial motion follows the lines of curved space inward around gravitational masses and this is correct so far as it goes. But why would a stationary apple released above the earth begin to move even if space there is curved? What is it about curved space that causes motion in the first place?

Gravitational attraction becomes easy to understand in the velocity density model. A gravitational field is a velocity density field that falls off by the square of the distance due to the geometry of 3-dimensional space. This means that the velocity density field is a velocity gradient field in which the intrinsic velocities at any point are greater in the direction of the gravitating mass. Since the intrinsic velocity of surrounding points is greater in the direction of the source this produces a velocity vector at every point that points toward the gravitating mass and objects in the field tend to move in the direction of the velocity vector.

Thus by understanding mass as spatial velocity we have a natural explanation for the fundamental nature of the gravitational force. Such an explanation was entirely lacking in Newtonian gravitation's attraction at a distance, and is still completely lacking in the curved space of general relativity. General relativity tells us that the presence of mass curves the surrounding spacetime but *offers no explanation at all for why* it curves it! In contrast, Universal Reality provides a clear and convincing answer for both the force of gravity *and* its relativistic effects. Masses actually are fields of velocity density gradients in space that inertial motion tends to follow, and because masses actually are fields of intrinsic spatial velocity they automatically slow time and produce other relativistic effects according to the STc Principle.

And conversely space itself consists of a field of velocity density. In flat space the density is the same everywhere so there are no resulting velocity vectors and inertial motion just traces a straight line. In the absence of gravitation in flat space there is no net velocity vector in any direction, but around masses there is a field of resulting velocity vectors pointing inwards towards the mass that defines inertial motion in that area. This is what a gravitational field actually is.

Apples fall towards the earth because of the velocity vectors produced by earth's velocity density field. And we stand on the earth because our inertial motion towards earth's center along its velocity vectors is blocked by its surface.

What we feel as the force of gravity is the intense fine vibration of the earth's mass increasing the velocity density of the spacetime around us, or equivalently the tug of the resulting velocity vectors on our bodies.

During the inertial motion of a free fall along the curve or velocity vectors of spacetime an observer experiences no force. It's only when that motion is interrupted and an observer stands on the surface of earth that he experiences a continual acceleration against his natural inertial motion. It is this feeling of resisting an inertial motion along a velocity density gradient in spacetime that is commonly but mistakenly called the force of gravity (Wikipedia, Equivalence principle).

Though the inertial trajectory of a falling object seems substantial from the point of view of an observer standing next to it on the surface of earth, the actual trajectory is not just through space, but through time as well. The distance traveled in *space* by a falling object is miniscule compared to the length of its world line through *time*.

For example the distance in space traveled in a 1.3 second fall to earth is a little over 24 feet, but the distance traveled in time is 1.3 light seconds, equal to the 240,000 mile distance to the moon! So the falling object's actual world line extends a distance through spacetime equal to that from the earth to the moon and is almost perfectly straight since it deviates only 24 feet in 240,000 miles, or one foot per ten thousand miles. Actually using a light beam to define a straight line the world line hasn't really curved, rather the spacetime it travels through is curved by that amount. So almost all of anything's spacetime velocity will usually be through time rather than space unless it begins to approach the speed of light.

In this example the slowing of the falling object's clock is too small to be measured. But in a stronger gravitational field it would become apparent. An observer falling with a clock experiences its proper time still ticking at c since his clock is falling along with him and is slowed by the same amount his internal biological clock is. Relative to himself and his own clock he's not moving so all his own motion appears to be through time.

Note there is a very slight difference in the time rate of a stationary clock in a gravitational field and a clock falling past it. Both experience the same slowing due to the gravitational field since they are both at the same position within the field. But the falling clock is also slowed slightly more due to its motion relative to the background of computational space. This second effect depends on the velocity of the falling clock as it passes the stationary clock. On earth the effect is negligible except for very fast moving mesons produced by cosmic rays whose half-lives increase due to their internal clock rates slowing.

To summarize, the mass of elementary particles consists of vibrational velocity, and the elemental computations these particles are involved in produces an entanglement network containing intrinsic vibrational velocities that are equivalent to the dilated spacetime of general relativity. It's this velocity density field that produces the slowing of clocks and the velocity density vectors that produce the gravitational attraction of objects towards the source of the field.

The velocity density model should also be consistent with frame dragging, tidal forces, and other relativistic gravitational effects. For the model to be accepted these effects must all be understood in terms of velocity density effects propagating across the entanglement network at the local speed of light, which is the rate at which all computational changes propagate across the entanglement network. Ultimately this model must correctly incorporate all the spacetime effects of the components of the stress-energy tensor in the Einstein field equations (Wikipedia, Einstein field equations).

Universal Reality's vibrational fields can be considered gravitational waves. Thus gravitation itself is actually gravitational waves produced as vibrations in spacetime by the presence of mass. However the recently detected gravitational waves predicted by general relativity are *changes* in these waves produced by rapid changes of position of extremely large masses (Wikipedia, Gravitational wave).

The velocity density model of gravitation holds for the other force charges as well. Each type of force can be modeled as a particular mode of vibrational velocity that forms a velocity density field centered on its charge(s). In particular positive and negative electromagnetic charges and poles take the form of fields of opposite helical rotations that reinforce or cancel each other depending on the direction of the twist as explained in the next chapter on Relativity and Electromagnetism. Because electromagnetism is also a form of energy it also produces gravitational effects in the same manner than mass does.

Because mass vibrations are such fine scale in place relative motions they appear the same in all (non-relativistic) frames and particles are said to have fixed rest masses. This is also true of the wave velocities of photons, which appear largely the same to all observers. This is in contrast to relative linear velocity, which clearly depends on even small observer velocities.

Though the velocity density around a single mass-energy particle is miniscule each additional particle adds its amplitude to the velocity density already present. So the cumulative effect around large collections of massive particles generates the extensive velocity densities in the surrounding space and clock time around stars and planets.

The vibrational velocity model of mass is consistent with the conservation of mass and energy as transformations of equivalent amounts of relative motion from one form to another. Otherwise there is no explanation of how or why mass-energy should be conserved. The conservation of mass and energy makes sense only if they are different forms of the same thing. Thus the conservation of mass-energy always involves the transformation of equivalent amounts of various forms of relative velocity from one form to another.

The beauty of our model of mass-energy as relative motion is that it explains both the conservation of mass-energy and gravitational time dilation and other relativistic effects. By the METc Principle mass-energy is relative motion whose value is its spatial velocity and by the STc Principle relative motion causes time dilation and other relativistic effects. Thus the vibrational velocity of mass increases the intrinsic velocity density of space around it and results in time dilation and other gravitational effects.

Thus a concentration of massive particles is a source of intense fine vibration that propagates outward through the surrounding entanglement network, and observers experience an intrinsic relative motion that slows their clocks in accordance with the STc Principle.

Thus if gravitation is understood in terms of the mass-energy that produces it, and mass-energy is recognized as a form of relative motion then the STc Principle becomes a truly unifying principle that describes both the time dilation of relative motion and gravitational time dilation as aspects of a single process.

So relative velocity produces gravitational time dilation just as it produces the time dilation associated with linear motion. In one case it's the kinetic energy of linear motion, in the other it's the vibrational mass-energy of elementary particles in aggregate. Both are just different forms of relative motion, which is why they are interchangeable and energy can be conserved through all its forms.

The vibrational relative velocity of mass and the relative velocity of linear motion are both forms of energy and each can be converted to the other, or to the relative wave motion of electromagnetic radiation. Particle interactions are little computational factories that convert one form of relative motion to another.

The fixed c value of total spacetime velocity is produced by a fixed number of processor cycles some of which are used up in computing spatial motion, leaving fewer to compute rates of temporal change. This automatically manifests as the STc Principle.

All forms of mass and energy are velocity in or of space; mass is space in relative vibrational motion to the velocity background of empty space. Space itself is the vibrational motion of the quantum vacuum as expressed in its zero-point energy value. On a large scale empty space is uniform relative motion, which can't be detected because there must be something to move relative to something else for motion to manifest. Thus particles must be created out of the quantum vacuum as particle component sets for there to be something to move relative to the background.

So space is just the presence of the energy of spatial motion, including the uniform lack of observable motion, and velocity is its measure. But motion takes time as well as space so spacetime is the presence of motion, of happening, and ultimately of life.

THE CLOCK POSTULATE

Universal Reality explains gravitational relativistic effects exclusively in terms of the total *velocity* at any point, the magnitude of the velocity being the combined linear velocity of an object plus the intrinsic velocity of the gravitational field being traversed. However some discussions of relativity assume *acceleration* is necessary to actually dilate time on the basis of Einstein's Equivalence Principle, which notes the equivalence of gravitation and an accelerating elevator in empty space (Wikipedia, Equivalence principle).

Though time dilation can be correctly analyzed in terms of acceleration to produce an equivalent result, the simplest approach is based on the rather ambiguously named 'clock postulate'. This states that the rate of a clock doesn't depend on its acceleration but only on its instantaneous velocity over all points on its world line. This was discussed in Einstein's original 1905 paper on special relativity as well as in subsequent kinematical derivations of the Lorentz transformations (Wikipedia, Twin paradox).

This means that we can ignore acceleration completely in the analysis of time dilation produced by spatial motion. Acceleration has no effect other than how velocity changes as a result. Thus we can modify the usual space traveling twin example to exclude acceleration altogether and still get the same result that the traveling twin's clock slows relative to the earth bound twin.

We can demonstrate this with a simple thought experiment by completely excluding acceleration from the twins example by adding a third traveler going in the opposite direction from the traveling twin who synchronizes his clock to the clock of the twin traveling away from earth as they pass. We find that when this third traveler returns to earth his clock will be slowed exactly the same as if the twin had instantaneously turned around and returned. Thus the acceleration of the twin turning around has been eliminated with no effect.

To complete the picture we can eliminate acceleration altogether if the outgoing clock doesn't take off from earth but simply synchronizes its clock with earth clocks as it passes at constant velocity to begin its journey, and finally by simply comparing the incoming third clock to earth clocks as it passes the earth without decelerating to land. In this example there is no acceleration of any of the three clocks, but the combined passage of time on the moving clocks is still less than on the stationary earth clock. The traveling twin (and his surrogate) has still aged less than the earth bound twin due entirely to non-accelerated velocity over their longer world lines.

So acceleration per se has no direct effect on time dilation. Its effect is due only to the fact that it changes spatial velocity. Thus even for an accelerating clock, time dilation can be correctly calculated just by integrating the changing instantaneous velocities all along the path of travel to get the total velocity along the world line.

By the STc Principle two clocks in relative motion will each *observe* the other clock running slower. By the STc Principle everything always has a constant combined velocity through space and time equal to the speed of light. Because part of the constant spacetime velocity of each clock is now seen as their relative motion through space, each sees the other's clock slow down to compensate for its increased relative velocity through space. The vector sum of velocity through space and through time is still c, but some of that velocity is now through space so less is through time.

This is an *observational* effect each observer sees in the other's clock since each is moving at the same speed relative to the other. The two observers each see the other's time slow by the same amount proportional to their relative speed. But once the relative velocity ceases and they are stationary with respect to each other this effect vanishes and their clocks are seen by both to be running at the same rate again.

However when the relative motion ceases there will still be a real and agreed difference in the amount of clock time that elapsed during the motion. This *actual*, as opposed to *observational*, effect depends on the extent of actual travel through space, on the actual spatial length of the respective world lines and the actual spatial velocities along those world lines.

So the unifying principle of relativistic time dilation is the STc Principle. The coordinate time of any clock is always *observed* as slowing from c proportional to its *relative* motion through space. However this effect is only *actual,* in the sense it's lasting and agreed among observers, if the motion was an actual motion through the computational background space as opposed to just an observed relative motion.

For example if clock A remains on the earth and clock B travels through space and returns they both see each other in relative motion during the trip with each other's clocks running slower, but when B returns and they compare clocks they both agree that only B's clock has actually slowed with respect to A's. This is because only B actually traveled in space and A didn't. (The motion of earth in its orbit can be ignored since it's miniscule compared to the world line of B in this example.)

Thus the key to understanding the relativistic slowing of time due to motion is that coordinate clocks always slow when in spatial motion, but it's only an actual permanent effect to the extent the spatial motion was actual motion through space as opposed to just motion relative to an observer who was moving himself. And acceleration has no intrinsic effect at all, the only thing that counts is the velocity of the spatial motion not whether it's inertial motion or varies due to acceleration.

This difference between actual and observational relative effects is what demonstrates there must be an absolute spacetime background relative to which actual spatial motion occurs. This can only be the actual computational space in which all motion is computed.

MASS VIBRATIONS & THE HIGGS FIELD

In our theory it's the processor cycles that compute the observable universe that convert numeric mass values into the fine vibrational velocities of mass. This means that the processor cycles are an analogue of the Higgs Field that physics suggests gives particles their masses (Wikipedia, Higgs boson). The processor cycles translate the numeric mass values of Level 0 into vibrational velocities as they are dimensionalized. Thus the processor cycles act as the Higgs field that gives particles their observational masses.

The amplitude of these vibrations is the gravitational strength of the particle mass. Thus more massive particles would have greater vibrational amplitudes than less massive particles. In the case of massless photons there are no intrinsic vibrations but instead extrinsic electromagnetic waves whose frequencies are their energies.

Gravitation is only additive and never cancels so the amplitudes of multiple particles can reinforce but never cancel. Thus they must be modeled as vibrational peaks in the elemental cells of space itself so they are all in the same position and can't cancel. And since the vibrational frequencies are the internal clocks of massive particles the frequencies of all additive vibrations of any given cell must all have the same frequency because any point cell in space can have only a single intrinsic space and time velocity ratio to as to obey the STc Principle. This explains why the intrinsic vibrations of mass reinforce but never cancel.

So the presence of additional masses simply increases the total velocity density of a field, which in turn produces the correct relativistic gravitational results. This effectively increases the density of spatial vibrations that must be traversed in any volume of space, which is equivalent to a greater spacetime curvature.

Thus it's the processor cycles that compute all processes that act as the Higgs field that converts numeric mass values into vibrational velocity densities and gives particles their observational masses. Since the fixed number of processor cycles determines the c value of the speed of light the value of c must be intimately related to the strength of the Higgs.

THE INCREASE OF MASS WITH VELOCITY

The processor cycles remaining after the computation of spatial velocities involved in an event go to computing the event's temporal velocity, its time dilation. The event's temporal velocity is the number of cycles used to compute the evolution of its internal details. This includes both the intrinsic clock rate of the particles themselves based on the presence of any internal vibrations of charges of the particles and by extension the rate at which particles interact.

Mass and other charges are modeled as spherical fields of spatial velocity densities. The respective charges are specific forms of vibrational velocity that produce the fields. Thus charged particles have internal detail in the form of their vibrations that is computed by the fixed allocation of processor cycles. Massive particles have the internal detail of their mass vibrations and as a result can never travel at the speed of light because there must always be some time allocated processor cycles to compute their internal vibrational detail.

In contrast massless photons and bosons have no mass and thus no internal detail to be computed by temporal processor cycles, and as a result time doesn't pass on their internal clocks. Photons have no internal proper time velocity and as a result all their velocity is through space and they always move at the speed of light through space. All the processor cycles that compute the evolution of photons are devoted to computing their spatial velocity since they have no internal details to be computed by any temporal cycles. In contrast their electromagnetic wave frequencies are computed externally rather than internally in the same sense linear velocity is.

Of course massless photons take time to move through space on the clocks of external observers but their commoving proper time clocks never advance. Time never passes on their own clocks, and as a result all their c velocity is through space. In contrast the necessity of computing the internal vibrational motion of the masses of massive particles prohibits them from ever moving at the speed of light.

Changes in fields including the gravitational field also propagate through the computational background at the speed of light since their *propagation* is a massless process that has no internal structure even though the fields themselves are fields of mass-energy that do have internal structure.

Thus our computational model consistently explains why massless photons always travel at the speed of light due to the allocation of processor cycles between computing velocity in space and velocity in time, while massive particles must always travel slower than the speed of light because they have internal temporal details that must be computed.

If we assume that mass at Level 0 where it's conserved is simply a numeric value, then it's the processor cycles that convert this numeric value into dimensional vibrations. By the MEv Principle, which states that mass-energy is spatial velocity, the amplitude of these vibrations, their velocity in space, is a particle's observational mass. Their amplitude is their intrinsic spatial velocity, their gravitational strength, and their frequency is their internal clock time rate.

By the STc Principle a particle's total spacetime c velocity is the vector sum of its time and space velocities, the vector sum of the frequency and amplitude of its mass vibrations. Thus any slowing of a particle's internal clock rate (vibrational frequency) will increase its observational mass (vibrational amplitude).

We can define a particle's *nominal rest mass* as the amplitude of its mass vibrations when it's at rest relative to an observer. The nominal rest mass is the rest mass noted in tables of particle data. In contrast a particle's *observational mass* is its rest mass when measured by an observer whether or not the particle is in motion relative to the observer. By special relativity the observational mass of a particle increases with its relative spatial velocity.

An isolated single particle at rest in empty flat space will have only a miniscule intrinsic spatial velocity in the amplitude of its vibrations and nearly all its vibrational energy will be expressed as a speed of light velocity through time in the frequency of its vibrations. In this case its observational mass will be its nominal rest mass.

However if the particle starts moving with spatial velocity this reduces its velocity in time. By the STc Principle linear velocity will significantly reduce the number of processor cycles available to compute the frequency of its vibrations and their frequency per P-time tick will decrease. Because the vector sum of frequency and amplitude must always equal c the observational rest mass (the amplitude of the particle's own intrinsic mass vibrations) increases and this is the source of the increase of mass with linear velocity predicted by special relativity.

When the particle is in a gravitational field the amplitude of its own intrinsic spatial velocity is added to the intrinsic spatial velocity of the field. The particle also 'feels' the total intrinsic velocity of the field it's in. By the STc Principle this further slows its own velocity in time, the frequency of its vibrations, in accordance with general relativity. This in turn increases its own intrinsic spatial velocity, the amplitude of its vibrations. So a massive particle in a gravitational field experiences a slowing of time due to gravitational time dilation and its observational mass also increases.

The increase in observational mass manifests as the increased *weight* of the particle in a strong gravitational field. This explains why masses in stronger gravitational fields are heavier because their weight, their observational mass, is increased by the slowing of their internal time velocity by the intrinsic spatial velocity of the field. Thus the STc Principle explains both gravitational time dilation and the weights of masses in gravitational fields as aspects of the same computational process.

The particle in the gravitational field experiences the intrinsic spatial velocity of the field. This reduces the frequency of its vibrations (its proper time velocity) which in turn increases their amplitude so that the vector sum of the particle's space and time velocities remains equal to c by the STc Principle. This is of course also true of all classical level objects, which are all composed of individual massive particles.

The frequency and amplitude of a particle's vibrations depends on the allocation of processor cycles used to compute them. In flat empty space with nearly all processor cycles used to compute velocity in time, the mass of a particle will be minuscule. However if the particle is in a gravitational field or gains linear spatial velocity this reduces the processor cycles used to compute its time velocity. This in turn makes more processor cycles available to compute the particle's own intrinsic spatial velocity increasing the amplitude of its vibrations which manifests as increased observational mass.

Thus the apparent or observational mass of a particle or object composed of particles increases when it experiences increased linear velocity or the increase intrinsic spatial velocity of a gravitational field. The increase in mass with linear velocity is an observed consequence of relativity (Wikipedia, Special relativity). So our model correctly explains the observed increase of mass with relativistic velocity as well as the nature of weight itself as aspects of the same computational process. The apparent increase in mass of an object with relativistic velocity is simply increasing its weight in the same way the weight of an object in a strong gravitational field is increased. The increase in mass of a rapidly moving object is an increase in *weight* and is exactly the same process as the weight of an object in a gravitational field. They are both examples of observational mass, which is what weight is.

All forms of mass-energy are relative motion (spatial velocity) of one type or another corresponding to the type of force charge involved. Because of zero-point energy resistance to relative motion with respect to the absolute background could be said to produce a resistance friction that manifests as mass. Relative linear velocity is converted to the relative intrinsic velocity of mass as the speed of light is approached.

All forms of mass and energy are forms of spatial velocity. Space itself is the minimal spatial velocity of the zero-point energy. Particle masses are elemental units of space (spatial velocity) crystalized around valid particle component sets. The mass particle component consists of spatial velocity in the form of very fine scale vibrations in the fabric of space. The amplitude of these vibrations is the particle's intrinsic spatial velocity, which is its observational mass. The frequency of these vibrations is the particle's internal clock time rate, its velocity in time. By the STc Principle the vector sum of the amplitudes and frequencies of these mass vibrations always equals the speed of light c.

Any additional spatial velocity a particle experiences reduces its velocity in time, its vibrational frequency, and increase the amplitude of its vibrations and thus increases its observational mass. This manifests as an increase in mass with linear velocity, and as the increase in observational mass that we call weight in a gravitational field.

DIMENSIONAL SPACETIME

The preceding discussion is more evidence that clock time and dimensional space are not a physical spacetime framework within which events occur, but are computed as the dimensional aspects of events themselves.

What we call spacetime is the logico-mathematical consistency among observer measurements. Human observers *interpret* this dimensional consistency as a preexisting empty physical spacetime within which things exist and events occur. However this is a projection of the logico-mathematical consistency of measurements into an apparently physical world constructed in our brain's internal simulation of reality. Physical spacetime is a reification of the logico-mathematical consistency of dimensional observations.

This logico-mathematical consistency is generated by the conservation of particle components in particle interaction events. Elemental particle interactions seem to be relatively simple logico-arithmetic computations that conserve the total amounts of particle components including dimensional particle components such as total mass-energy and spin orientation. Particle component conservation entangles the particles emitted by events on their conserved particle components, meaning that the values of particle components of emitted particles bear fixed relationships to each other.

Sequences of events form entanglement chains and all events together form an entanglement network of particle components in which the values of all particle components in the universe are related across particles. It's the internal dimensional consistency of this entanglement network that human observers interpret as an enveloping spacetime container for events.

For events to exactly conserve particle component values of energy and momentum the dimensional fuzziness of particles must be decohered to exact (within limits of Uncertainty) dimensional values by events so they can be exactly apportioned among emitted particles.

These exact dimensional and other particle component values constitute mutual measurements of particle observables. It's the observational results of these measurements that are the observable data of the universe. The actual computational universe and its values are unobservable. The observable universe exists only as observer compilations of observed values ultimately deriving from decoherences and particle component values.

OBSERVER FRAMES

All observable values are measurements relative to some observer frame. The relativistic equations describing these frame views have the same covariant forms as they do with respect to the preferred universal background frame in which they are actually computed. The difference is that the frame they are actually computed in becomes the preferred universal frame with respect to which actual rotation and actual world lines are relative to.

To the extent that observer frames are not aligned with the actual background frame, their relativistic effects are observational rather than actual and vanish without lasting effect as soon as any relative motion ceases. The equations have the same general relativistic form, but the effects are observational rather than actual. And of course observers don't use the same number system, scales or units in which the universe is actually computed but which remain unobservable.

'Physical' spacetime is the logico-mathematical consistency among observations projected into the simulated universe produced by an observer mind, and more generally the shared simulation of a group of like-minded observers such as humans or physicists.

Logico-mathematical consistency means that particle observations obey the predictions of general relativity at large scales and quantum theory at small scales. However the various interpretations of each theory, all based on a pre-existing spacetime within which events occur, are outmoded and must be replaced. The computational interpretation of Universal Reality in which both theories can be united is much more fruitful, reproducing general relativity at large scales and quantum theory at small scales.

Another major difference between the actual computational frame in which the universe is computed and observer views is that observers invariably incorporate past dimensional observations into their dimensional simulations whereas the actual computational frame of the observable universe includes only current values as they are computed.

Because observers are part of the universe they and their dimensional views are all ultimately computed by the elemental computations so all observer views are consistent aspects of the actual computational universe. Observers are part of the entanglement network and thus part of its numeric consistency. The elemental program that computes the consistency of the whole network also computes the observers that exist within it and the relativistic effects they observe.

Though human observers are all emergent programs that view the entanglement network through their simulations of it, relativistic observer views can be analyzed in terms of imaginary single point observers and frames if we are careful. This is fine because all observations ultimately reduce to single particle events though they are typically mediated by chains of events through laboratory instruments that scale results up to the classical level, or through the even more complex perceptual systems of observers.

It must be remembered that the numeric consistency of the entanglement network is only established through the elemental Level 0 computations that generate the numeric values and their relationships. The numeric consistency is a matter of the equations of the quantum vacuum consistently computing exact numeric values from previous ones.

The quantum vacuum also computes observers as part of the entanglement network so observers are part of its numeric consistency. So the view of the entanglement network that any observer sees is also computed by the elemental program of the quantum vacuum.

The same equations that actually compute reality in its own background frame are also used to compute views of the entanglement network from moving frames within it. The relative motions, orientations, and coordinate origin points of their frames define individual observers.

Because observers view the entanglement network from the perspective of their own frames they tend to ascribe their own relative motion to the entanglement network. Thus the elemental program computes the relativistic views of observables as if they had the inverse of the observer's own relative motion.

Because this is all calculated by reality it's the real actual view of the observer, but it's not a view shared by other observers with different relative motions. Every observer with a different relative motion will have a different view of reality computed by the quantum vacuum that seems correct to him but isn't shared by other observers.

All frames in relative motion will have different relativistic views and these are the real actual views of observers from those frames. However only relative motion with respect to the background frame in which the entanglement network is computed has actual lasting relativistic effects. All other relative motion produces relativistic effects that are observational though they are quite real and genuine in the frames of observers. This means that when observers meet and compare relativistic effects they aren't lasting or agreed unless they were produced by relative motion with respect to the actual background frame in which everything is computed.

For example observers in relative motion will each see the other's clock run slower. But when they meet only the clock(s) that moved with respect to the entanglement network background will show less actual elapsed time. And both observers will then agree on this.

Similarly rotational motion will only produce actual centrifugal effects to the extent it's with respect to the background frame of the entanglement network. Take an observer at the center of a merry-go-round in deep space. Visually he has no way to know if the merry-go-round is rotating or not because there is no visual background. However if the merry-go-round is rotating with respect to the entanglement network loose objects tend to fly off it whether or not the observer is rotating with it or not. So the view of the observer is observational but the effects with respect to the actual computational background frame are actual. The effect depends entirely on the rotation with respect to the computational background and not with respect to the observer.

The relative motion of observers can compute identical *observational* effects but these are local and change with the observer's relative motion. The absolute computational background is necessary to maintain the logico-mathematical consistency of the universe and thus its existence.

Educated observers are able to discover the equations reality uses to compute a relativistic entanglement network and invent the laws of general relativity to convert among the views of observers but this occurs at the emergent level of their simulations of reality.

Ultimately the quantum vacuum even computes the simulations of observers as emergent aspects of the entanglement network, but at this point it makes sense to consider the simulations themselves as emergent programs that perform the computations.

Observer views are all part of the super consistency of the universe, the fact that the universe is logico-mathematically consistent across all observer views at all levels of emergence, and this is all due to the super consistency implicit in the design of the complete fine-tuning.

The simulation programs of observers are imperfect models of the computational structure of the universe. They selectively filter information and fragments of its logico-mathematical structure and apply them to modeling situations of importance to the individual observer. Because they are based on small amounts of highly filtered input data their computational results are inexact but good enough in general for the observer to function and survive as part of the actual entanglement network that computes it.

The simulation programs do have significant adaptive advantages in that they are able to store and compare past data states to infer causality and thus predict future events, and they are able to compute in terms of individual things and events and relationships extracted from vast floods of raw data. Even though not exact and often inconsistent from thought to thought this program is highly adaptive and easily switches among small-scale situational models to compute reasonably effective actions. It enables on the fly comprehension by redefining things as it goes, something the quantum vacuum is unable to do at the level at which it actually computes the entanglement network. Nevertheless it's the entanglement network that ultimately enables all this through the super consistency of its complete fine-tuning.

A NEW DARK MATTER THEORY

The existence of an invisible form of matter called dark matter was first proposed to explain observational anomalies in the motion of galaxies. For example observations suggest that galaxies rotate as if they had halos of invisible mass around them because they are rotating faster than would be expected based on their apparent masses. The amount of dark matter necessary to explain the movements of galaxies is huge, about 5 times the amount of visible matter in the universe (Wikipedia, Dark matter).

Dark matter has been sought in the form of various types of new particles but so far none have been found. However Universal Reality suggests another possible explanation for the dark matter effect, which so far as we know is original to the author's 2013 book 'Reality'. This proposal is a simple and rather obvious consequence of the Hubble expansion.

The Hubble expansion is an expansion of the relatively empty space *between* galaxies and galaxy clusters which makes up most of the universe. By contrast the space *within* galaxies isn't expanding because it's gravitationally bound by their mass (Misner, et al, 1973, p. 719). Thus the earth, the solar system, our galaxy, and we are not expanding but the space between galaxies is expanding.

The result is an uneven Hubble expansion that warps space around the boundaries of galaxies; precisely in the area that dark matter is expected to be found! And from general relativity we know that any warping of space must manifest as a gravitational effect. Thus we have a natural explanatory mechanism for the dark matter effect that involves only the expected warping of space from the uneven Hubble expansion around galaxies and doesn't require the existence of any new particles.

This warping may or may not be the cause of the entire dark matter effect, but it certainly should be producing a large gravitational effect, since the uneven expansion over the lifetime of the universe should produce a very large warping of space.

Distributions of dark matter can be mapped by tracing gravitational deviations of the expected paths of light beams from sources beyond them. These maps indicate a distribution of dark matter generally around galaxies but sometimes offset as well. However there is nothing to prevent these Hubble space warps, once they are created, to have a life and movement of their own. Thus dark matter distributions should initially form as halos around galaxies and galaxy clusters but then be able to move as massive objects on their own.

Once Hubble warps are formed they are effectively just additional areas of gravitational mass that can move through space just as galactic masses do. The continued existence of a dark matter mass is not dependent on the original galaxy it was created from. There will be a continuous creation of new dark matter warps around galaxies, but once created these can trail away and should leave detectible plumes of warping behind that indicate how galaxies moved over time.

Over the course of the expansion of the universe the actual effects will be extremely complex because the distribution of galactic matter with time is extremely complex. It should be fairly easy to test at least the viability of this theory by comparing the current distributions of dark and visible matter and inferring their relative motions over time and making a calculation of whether the expected warping would account for the gravitational effects of known dark matter concentrations.

This is one possible explanation of the dark matter effect, but not necessarily the only one. Nevertheless there should be a very substantial warping due to the uneven Hubble expansion, and that warping should be producing quite a large gravitational effect. Where is that effect if it isn't the dark matter effect? It must be somewhere. The evidence seems quite strong and it certainly simplifies things by not requiring any new unknown types of particles.

This theory of dark matter also neatly explains why dark matter is dark. Not being an actual form of particulate matter it obviously doesn't emit light. Thus it's invisible and interacts with regular matter only via the gravitational force.

RELATIVITY & ELECTROMAGNETISM

ELECTROMAGNETISM

The electromagnetic force is quite interesting because the relationship between its electric and magnetic components behaves much as energy, space and time do. Just as energy is space in relative motion, so magnetism is electricity in relative motion. Electricity and magnetism are two orthogonal (90°) components of a single underlying entity just as space and time are, and in both cases each is transformed into the other via relative motion.

In physics, a magnetic field is the relativistic part of an electric field, as Einstein explained in his 1905 paper on special relativity. When an electric charge is moving from the perspective of an observer, the electric field of this charge due to space contraction is no longer seen by the observer as spherically symmetric due to relativistic shortening along the axis of motion, and must be computed using the Lorentz transformations. One of the products of these transformations is the part of the electric field that only acts on moving charges which is called the magnetic field (Wikipedia, Electromagnetism portal).

This similarity between electricity and magnetism and space and time underlies the Kaluza-Klein Theory in which electromagnetism is modeled as a 5th compacted dimension and an electric charge is a standing velocity in that 5th dimension (Halpern, 2006). The beauty of this theory is that when this 5th dimension is added to the 4 dimensions of general relativity, Maxwell's equations of electromagnetism automatically emerge (Wikipedia, Kaluza-Klein theory).

Thus one can consider electricity as a fundamental force and magnetism as electric charge(s) in motion. This motion can take several forms. At the elemental level of particle components all electrically charged particles have an intrinsic half integer spin, which effectively rotates the charge about an axis.

Because spin gives its associated electric charge rotational motion spin manifests as magnetism and particle spin is the intrinsic underlying unit of magnetism. Since spin about an axis produces an orientation of the axis spin is equivalently an intrinsic underlying unit of dimensional *orientation* and angular momentum relative to the computational background. This is critical to understanding Newton's Bucket as we have seen.

The quantum mechanical velocity of electrons in atoms produces the magnetism of permanent ferromagnets. Ferromagnetism is due primarily to the alignment of the spins of ionic (outermost orbital) electrons in atoms. In most materials the spins of particles are randomly aligned and tend to maintain their random alignments just as spinning gyroscopes do.

Materials made of atoms with filled electron shells have a total dipole moment of zero, because every electron's magnetic moment is cancelled by the opposite moment of the second electron in the pair (see the section on Bound Entanglement in the next chapter for how this works). Only atoms with partially filled shells (i.e., unpaired spins) can have a net magnetic moment, so ferromagnetism only occurs in materials with partially filled shells.

These unpaired dipoles (often called "spins" even though they also generally include angular momentum) tend to align in parallel to an external magnetic field, an effect called paramagnetism. Ferromagnetism involves an additional phenomenon; the dipoles tend to align spontaneously, giving rise to a spontaneous magnetization, even in the absence of an applied field (Wikipedia, Ferromagnetism).

A fundamental characteristic of magnetism is because it's due to rotational spin about an axis and the poles of the spin axis are spinning in opposite directions from the point of view of the exterior, magnets always appear to have equal and opposite magnetic poles. Because magnetism is fundamentally a product of axial rotation there can be no isolated magnetic monopoles. Magnetism is always dipole.

However magnetic poles are actually an illusion because magnetic field lines continue through the interior of a magnet and just emerge at the other pole in the opposite direction. Thus magnetic field lines always form closed loops as opposed to electric field lines, which radiate outward from electric charges. And magnetic poles are simply a name given to where a denser concentration of field lines enters or exits a magnet.

So the magnetic force is actually along the field lines proportional to their density, which is greater at the 'poles' of a magnet. Thus it appears the poles are doing the attracting or repulsing but it's actually the density of field lines themselves. The opposite poles are due to the field lines pointing inward as they enter at one pole and pointing outward as they exit at the other.

Thus magnetism doesn't really have positive and negative poles. It's just a matter of which direction the lines of force are pointing and how dense they are. So for example the magnetic field around a straight current carrying wire has *no poles* because the field lines are all circularly concentric around the wire. So it's the density gradient of the lines that is greater towards the wire that exerts a magnetic force either towards or away from the wire.

Magnetism is different from electricity in this respect, which does always come in positive or negative charges. And electric charges are always isolated to individual particles.

Again with electricity other charges are attracted or repelled not so much by the charges themselves but because the electric field is a velocity density gradient in spacetime with a velocity vector at every point. As with mass the electromagnetic field is an inseparable part of the actual charge and other charges tend to move along velocity vectors in the field gradient.

The second form of magnetism due to the movement of electric charges is due to the orbital motion of electrons in atoms. This orbital motion produces quite a strong magnetic force but like the spins of most particles it's randomly oriented among atoms and mostly cancels out.

The third form of magnetism is due to the motion of electric charges in currents. When electric charges move through a wire they generate a magnetic field encircling the wire according to the Right Hand grasp rule. When many wires are wrapped tightly in a coil (a solenoid) the magnetic field generated within the coil is multiplied and when the current is properly modulated will rotate an iron rotor. This is of course the principle of the electric motor.

The magnetism generated by particle spin can be easily understood by analogy to that generated by a moving current in a wire. If we slice the spinning particle open on one side and lay it out flat we see that the magnetic field encircles the particle just as it does a wire.

There is also an opposite effect in which a changing magnetic field produces an electrical current as in a generator. The principle, Faraday's law, is that an electromotive force is generated in an electrical conductor that encircles a varying magnetic flux. Motors and generators are similar in form and many motors can be mechanically driven to generate electricity and frequently make acceptable generators.

So all three of these magnetic effects are manifestations of electric charges in motion. Magnetism is electric charge in relative motion and this relative motion can be either that of the electric charges themselves or of an observer relative to them. Thus magnetism is a clear everyday example of relativity in action. Whenever we experience magnetism we actually experience relativity in action.

An observer at rest with respect to a system of static free electric charges will see no magnetic field. However if either the charges *or* the observer begins to move the observer perceives it as a current and an associated magnetic field. A magnetic field is simply an electric field seen in a moving coordinate system. It doesn't matter whether the electric field or the observer is moving; all that counts is their relative motion. How this works as an effect of Lorentz contraction along the direction of motion is depicted graphically at (Schroeder, 1999).

However recall that actual relativistic motion is with respect to the computational space in which it's computed, as opposed to observational relativistic motion, which is simply relative motion between observer and what is observed. So there will be an absolute transformation of electric to magnetic force from actual motion with respect to computational space, but only observational effects in the frame of observers due to their own relative motion with respect to computational space.

So magnetism is actually a relativistic effect of electricity, and electricity is transformed into magnetism by relative motion just as velocity in time is transformed into the mass-energy of velocity in space by relative motion. Both are examples of the Lorentz transform, which is simply the Pythagorean theorem describing the projections of a single vector onto orthogonal coordinate axes. Thus the electric and magnetic fields are 90° orthogonal projections of a single underlying entity just as space and time velocity are.

Thus when we play with a magnet and observe its effects we should realize it works because of the enormous in place velocities of its electric charges at the particle and atomic levels. The energy within matter is enormous and thus the in place velocities are enormous. It's only the near exact balance of forces that holds the energy of particles together into the seemingly ordinary and trivial objects around us.

THE ELECTROMAGNETIC FIELD

Though electric and magnetic fields are usually considered separate but related entities that can produce each other, there is actually only a single integrated electromagnetic field that is best understood in terms of the electromagnetic tensor (Wikipedia, Electromagnetic tensor). This tensor describes the relationships between the spatial vectors of the electric and magnetic fields.

In the electromagnetic tensor the individual electric and magnetic fields change with the choice of the reference frame, while the tensor itself doesn't. The electromagnetic tensor takes the form of a 4 x 4 matrix in any particular coordinate basis where the components of all but the diagonal cells are the values of the magnetic field along each coordinate axis and the values of the electric field divided by c along each coordinate axis. Thus there are 6 independent components of the tensor; E_x, E_y, E_z (the electric field) and B_x, B_y, and B_z (the magnetic field).

Thus every point in the field has a vector for E and for B along all three coordinate axes in whatever coordinate basis is being used. The values can be either plus, minus or zero for each vector component.

When the electromagnetic tensor is multiplied by a metric tensor expressing a change in coordinate basis, such as relative motion to an observer, the individual B and E values change. The form of the tensor gives the correct transformations of electricity into magnetic fields and vice versa for any observer frame, though there's still only a single electromagnetic field manifesting as a combined electric and magnetic field depending on the frame of reference of the observer.

This is expressed by the fact that the tensor has an invariant that doesn't change with transformations of coordinate basis. This invariant is $\mathbf{B}^2-\mathbf{E}^2/c^2$, which means that the total electromagnetic force is conserved in all coordinate transforms no matter how the individual electric and magnetic forces transform into each other (Wikipedia, Electromagnetic tensor). This is exactly analogous to the conserved transformation of space and clock time velocities into each other expressed by the STc Principle.

The solution to the tensor gives the force vectors of the two fields at every point and the total of all force vectors for all points traces the combined lines of force for both the electric and magnetic forces.

The electromagnetic tensor has another invariant 4/c (**B•E**), the dot product of the magnetic and electric force vectors, which roughly means that electricity and magnetism are orthogonal manifestations of a single electromagnetic force.

The invariance of the space-time four-vector is associated with the fact that the speed of light is a constant. The invariance of the energy-momentum four-vector is associated with the fact that the rest mass of a particle is invariant under coordinate transformations and the invariance of the electric-magnetic four-vector is associated with the fact that the total electromagnetic field is invariant under coordinate transformations. In these invariances we see fundamental principles of the universe at work.

Thus time turns into space with increasing velocity, and electricity turns into magnetism with increasing velocity. And increasing spatial velocity is increasing mass-energy. Thus increasing mass-energy turns time into space and electricity into magnetism.

Now if we just take the lines of forces as consisting of helices in accordance with our representation of field energy as various forms of intrinsic velocity in space we get a proper attraction and repulsion model of plus and minus poles and charges of these two distinct but interrelated forces.

THE HELICAL FIELD MODEL

Electromagnetism is another form of energy and thus according to Universal Reality a form of relative motion, with a strength equal to its velocity density. Like mass electric charges are modeled as spherical fields of velocity densities in spacetime radiating from the center of the charge. These fields of spacetime distortion alter the proportion of space and time distances and velocities at points in the field and the field gradient produces velocity vectors that induce relative inertial motion in other particles.

Since charged particles also have mass they are associations of two kinds of velocity density, one produced by their mass, and another by their charge. The fields are easily visualized as spherical areas within a *flat Euclidean* space in which the relative distances and velocities of time and space are shifted at every point in the field as described in the previous chapter.

Both gravitational and electromagnetic fields fall off as the square of the distance due to the simple fact that in 3-dimensional space the area of the surface of a sphere increases by the square of the radius. Thus the strength of fields falls off inversely with distance. Thus the constant strength of the field is simply diluted by the increasing volume of 3-dimensional space as the distance from the center increases.

This model of mass and gravitation suggests a similar model for electromagnetism. This is a very neat theory original to Universal Reality with a lot of explanatory power. It also provides an excellent explanation for how the standard theory of electromagnetic fields as virtual photons works.

The difference in the vibrations of mass and electric charge is in the form of the vibrations. Electric charges are spherical fields of *helical* spacetime distortions in the surrounding dimensional fabric. In other words charges produce a field of miniscule corkscrew twists in the surrounding spacetime that form the field lines of both electric and magnetic fields and increase the velocity density of points in space. Of course the actual fields are continuous and fill all space, the field lines are just a graphical sampling of the entire actual field.

These spacetime distortions produce velocity vectors felt mainly by other charged particles whose own fields couple to them since their helical distortions tend to reinforce or cancel each other out depending on their direction of twist. Electric and magnetic field lines are modeled as separate orthogonal projections of these helical distortions in space.

The transformation of electric force into magnetic force with the spatial velocity of charges occurs as the helical vortices of the electric field begin to tilt orthogonally into the helical vortices of the magnetic field according to the right hand grasp rule. With greater and greater velocity (current flow) the helices tilt more and more and become a magnetic field that appears as field lines of magnetism perpendicular to those of the electric field lines.

So current velocity transforms electricity into magnetism by flipping the helical vortices of the electric field into the orthogonal direction where they become helical vortices of the magnetic field. If the actual velocity of the charges approached the speed of light all the electric field vortices would flip over into magnetic field vortices and the electric field would become entirely a magnetic field due to the conservation of the total electromagnetic field.

These helical spacetime distortions are generated by individual charges and rotate in two possible directions, either in the clockwise or counterclockwise direction. These correspond to positive and negative electromagnetic charges or positive and negative magnetic poles. The fact there are only two possible rotational directions for helices neatly explains why there are only two electromagnetic charges and two magnetic poles.

The effective diameter and density of twists of these helices is fixed since the charges or spins generating them are the fixed plus or minus electrical charges and spins of elementary particles. The densities are additively scaled to produce the measured values of electric and magnetic forces produced by multiple particles.

The spins of the elementary electromagnetic charged particles clearly produce the helices of their electromagnetic fields, which extend outward from the spinning charges.

While the vibrations of mass come in different amplitudes and frequencies due to the non-proportionality of particle masses and the additive nature of the gravitational force, the helical vortices of electromagnetism have identical forms because the strength of their elemental charges are identical. Presumably multiple charges just add to the number of identical helices to produce a denser field. This enables the helices of electromagnetic fields to cancel or reinforce depending on whether they are turning in the same or opposite directions.

The relativistic effect of linear motion on mass vibrations is a tilting of time velocity into space velocity while the relativistic effect of linear velocity on the electromagnetic helices tilts the individual helices from parallel to the electric field lines towards the perpendicular magnetic field line orientation. The individual helices don't move they just tilt in place. Their individual projections on the electric field lines is the electric force and their perpendicular projections are the magnetic force and establish magnetic field lines.

So the electric force is analogous to velocity in time and the magnetic force is analogous to velocity in space. Both velocity in time and the electric force tilt into their alter egos with linear velocity and the vector sum of each with its alter ego is conserved. This similarity in form is why the electromagnetic force can be modeled as a compacted 5th dimension in Kaluza-Klein theory though our helical model is preferable.

If the charge producing the field of helices is set in motion relative to an observer the individual helices begin to tilt from parallel to the lines of electric force towards perpendicular to them. The perpendicular projection of the tilted helices becomes the magnetic field and the parallel projection to the lines of force is the electric field. This is the relativistic source of the magnetic force and its field lines.

Helices cancel each other out when they are rotating in opposite directions and reinforce when they are rotating in the same direction. Thus helices rotating in opposite directions cancel where they are pointing in the same direction, and reinforce where they are pointing in opposite directions. And helices rotating in the same direction cancel when they are pointing in opposite directions, and reinforce when they are pointing in the same direction. This is the key to understanding magnetic attraction and repulsion, and the repulsion and attraction of electric charges.

Thus in areas *between* separate poles or charges of the *same* sign the helices cancel each other out and they reinforce in areas *outside* the charges or magnetic poles. Thus the velocity density of spacetime is increased in the areas outside and unaffected in between. Thus the velocity vectors at the points of the poles or charges are directed away from each other and this is the source of the *repulsion* of identical magnetic poles and identical electric charges.

And in areas between separate *opposite* poles or charges the helices reinforce and they cancel beyond producing an area of strong velocity density between the charges so the velocity vectors where the charges are located point towards each other and this is the source of the *attraction* between opposite charges and opposite poles.

In areas external to two opposite charges or poles their field helices will be rotating in opposite directions and will almost completely cancel each other out. This is why there is no net magnetism in most materials because their collective helices cancel each other out in external areas. Only in materials where slight spatial imbalances of charge polarity exist and are aligned will external helices not be completely canceled. In such cases magnetic effects will be present and the materials will be magnetic.

Thus Universal Reality suggests a simple and elegant new model of electromagnetic attraction and repulsion in terms of the spacetime dilation generated by a helical velocity density just as it did for the similar gravitational effects of mass in the form of simpler vibrations. And the specific forms of the spacetime distortions produced by mass and electric charges neatly explain why there are two opposite electromagnetic charges and only a single gravitational charge.

Both theories are explained in terms of the same spacetime velocity density model, and both are complementary distortions that can exist together in the same spacetime volumes, which are observer views of the computational data structure of the entanglement network.

Thus both mass and charge can produce their different distortions in spacetime simultaneously. They work together naturally in standard 4-dimensional spacetime. Electromagnetic effects propagate across the curved spacetime of general relativity, and gravitational effects are produced by a simpler form of velocity density. Gravitation and electromagnetism are simultaneous distortions or velocity densities in spacetime of different forms. But why the intrinsic velocity of a gravitational field doesn't also tilt the electric force into magnetic force is an open question.

As forms of energy both mass and electromagnetic charges increase the velocity density of spacetime in their specific ways and this affects the relativistic behavior of objects within spacetime but most of the velocity density of electromagnetism consists of helical distortions that largely cancel each other out but also strongly couple to the helical fields of other charges.

Thus Universal Reality produces a simple and elegant theory of both mass and gravitation, and of electric charges and magnetism in terms of different forms of relative motion that both work according to the same underlying MEv Principle of all types of mass-energy as forms of spatial velocity.

And this model is quite easy to visualize and understand in terms of a flat Euclidean spacetime containing fields of the two forms of velocity density that produce velocity vectors in the direction of slower time. This Neo-Euclidean model is equivalent to the curved spacetime of general relativity, but much easier to comprehend because it directly reflects our actual flat Cartesian view of spacetime.

Think of each point in this Euclidean spacetime having a fixed c value (speed of light) of combined time and space velocities. The relative motion of either mass or charge just distorts the proportion of time and space velocities so that time slows and distance lengthens. We can still think of the overall spacetime as Euclidean but the relative velocity density of space and time at every point is distorted by mass or electromagnetic charge.

Both the vibrational distortions of mass and the helical distortions of electromagnetism produce spacetime distortions proportional to the relative velocity of their energy. Thus both produce velocity vectors that determine the inertial motion of test particles in their fields that are different depending on whether or not the test particle is charged.

Gravitation is a much weaker force than magnetism, the strength of the relative motion produced is 10^{-36} times weaker than that produced by electric charges, and thus the field densities and vector velocities are much less for a particle of mass than a unit of charge.

However because mass has only one charge (there are no negative masses) masses are all attracted to each other and tend to clump without limit into planets, stars and galaxies and produce very large gravitational velocity density fields.

On the other hand equal electric charges repel each other and cannot clump (except in very small units under the influence of the strong force in nuclei). Opposite charges do clump at a distance in the form of atoms and molecules, but their opposite helical spacetime distortions almost entirely cancel each other out beyond the clumps when they do, so most of the large scale structure of the universe is due to gravitational velocity density fields.

Thus though electromagnetism is intrinsically far stronger than gravitation, gravitation rules on cosmic scales and electromagnetism mainly just holds atomic matter together with external effects largely cancelling out. There are large magnetic fields on cosmological scales but their strength is generally much less than gravitational fields.

As a form of energy the helical spacetime distortions of electric charges do have some gravitational effect on uncharged particles as predicted by the stress-energy tensor of the Einstein field equations, but these are generally negligible compared to those of mass. Since charged particles are normally paired the helices largely cancel in areas external to the particles and the velocity density between closely adjacent points will be minimal thus the velocity vector will be relatively small compared to the force on a coupled charged particle which is effectively canceled in one direction and doubled in the other.

This is how our velocity density model addresses the initially vexing question of why the gravitational force is so much weaker than the electromagnetic force but curves space so much more.

The gravitational force is due to the vanishingly *small difference* in velocity density on either side of each particle in a field along an axis towards its source mass. Thus the resulting velocity vector towards the gravitating mass is also extremely small. However the helical velocity densities of the electromagnetic force couple to those of other charges and so completely cancel on one side and completely reinforce on the other along their common axis. This effectively doubles the *entire value* of the velocity density at the point location of the particles. The difference in gravitation's small *gradient* in velocity density at the particle scale and electromagnetism's doubling of the *entire* velocity density is enormous. Thus the electromagnetic force is much stronger than the gravitational force.

The velocity density produced by electromagnetism is considerable, but since charges are normally paired ambient velocity densities effectively cancel and the difference on proximate and opposite sides of *uncharged* particles is effectively nonexistent so gravitational effects will be vanishingly small. Note also that mass produces a much greater gravitational effect than energy since by $E=mc^2$ the equivalent amount of mass m in a unit of energy is E/c^2 which is an extremely small number.

Another difference between electromagnetic and gravitational fields is that electromagnetic fields can be shielded but gravitational fields can't be. Again this is due to the fact that the helical velocity densities of electromagnetism come in two opposing rotations. Thus it's generally possible to construct a shield that either damps or diverts an electromagnetic field but adding any shield made of mass or energy just adds to a gravitational field rather than blocking it.

Thus our model of electromagnetism seems reasonably consistent with standard scientific theory and explains its basic concepts well in terms of velocity densities as it also does mass and gravitation and the conservation of mass-energy.

PHOTONS

Photons of electromagnetic radiation are now easily understood as units or quanta of the electromagnetic field that break away from the charges that generate the field and fly off on their own taking some of an electron's orbital energy with them. They can also be absorbed by orbital electrons and kick them into higher energy orbitals by increasing their energies. Photons are essentially free quanta of electromagnetic fields that either break away or are absorbed back into fields in the process of emission or absorption of orbital energy.

Thus ejected photons are units of orbital velocity converted into the helical waveform of the electromagnetic field. This is an excellent example of how the conservation of mass-energy always involves the conversion of one form of spatial velocity into another.

The emission of photons of electromagnetic radiation is the result of the acceleration of an electric charge(s). This can occur either due to a free electrical charge or system of charges changing direction as in an antenna, or when an electron decelerates to a lower orbital and emits its lost orbital energy as a photon of electromagnetic energy, or if an electron absorbs a photon which accelerates it into a higher orbital.

Because they are quanta of helical electromagnetic fields photons can be described as helical vortices in space just as electromagnetic fields are, but in the form of localized packets of the field moving through space at the speed of light. The helical packets of photons are no longer attached to the particle generating the field and thus fly off at the speed of light. As explained previously photons always travel at the speed of light by the STc Principle since they have no internal structures to be computed and thus no internal proper clock time velocity.

So long as a mass or charge is present its velocity density effects propagate across the entanglement network (spacetime) at the speed of light. This is because all computational effects propagate through the entanglement network at the speed of light. Remove the mass or the charge and the surrounding velocity density field dies off at the speed of light.

Thus if an individual helix breaks away from a charge it will naturally propagate through spacetime at the speed of light. Photons of light, and other forms of electromagnetic radiation, are just individual electromagnetic helices freely propagating through space at the speed of light because they aren't tied to a source charge. This typically occurs when electrons transition to lower orbital energies and emit the excess energy in the form of a photon.

Photons are actual individual helical distortion packets moving linearly through spacetime as opposed to fields of helical velocity densities in all directions around charges. They carry energy proportional to the frequency of their helical rotations. By contrast the helical spacetime distortions around charges can be considered to be composed of virtual photons, as they are not transmissions of energy but do produce energetic effects when interacted with. They both consist of helical waves of the same basic form though the electromagnetic waves of photons are considered to be orthogonal combinations of electric and magnetic waves. Together these oppositely oscillating waves result in a helical spacetime distortion traveling at the speed of light.

Thus Universal Reality naturally explains light as a form of electromagnetic energy and naturally explains why it moves at the speed of light through the spacetime entanglement network. Again though, the speed of light is actually the speed of clock time, the speed at which computational effects propagate through the dimensional entanglement network that humans interpret as spacetime. Photons of electromagnetic radiation are fundamentally computational effects in this sense.

The helical distortions in spacetime directed outward from charges have fixed amplitudes and frequencies and wavelengths since the basic units of electrical charge and spin that produce them have fixed strengths. They differ only in the direction of their helical twists corresponding to the sign of the charge or pole that produces them.

However the electromagnetic radiation of light is not anchored to a fixed charge so its helices can be produced in a more or less continuous spectrum of frequencies proportional to the amount of relative motion converted to produce them. Thus the energies of photons range from radio waves to visible light through gamma rays.

Like the helices extending from charges, those of electromagnetic radiation can also rotate in either a clockwise or counterclockwise direction. This accounts for the possible clockwise and counter clockwise circular polarizations of light. In most cases, such as the light from the sun, light beams are a mixture of the two polarizations.

The helical waves of photons generally don't cancel each other out or cause attraction or repulsion because they tend to be mixtures of many different frequencies and due to their great velocities effects tend to be more or less instantaneous and immediately cease. However coherent beams of photons of the same frequency such as those emitted by lasers can interfere if correctly tuned (Wikipedia, Laser).

Like all relative motion these helical waves are dimensional aspects of the entanglement network and thus extend perfectly well across the vibrational fields produced by mass. Thus they are naturally subject to relativistic effects. They follow and add to the lines of spacetime curvature produced by the presence of mass. So both models together appear to correctly model electromagnetic fields in the curved spacetime of general relativity.

ELECTRICITY

It's common knowledge that opposite electrical charges attract and identical charges repel each other. This is called electrostatic force and is experienced at the classical level in the static attraction and repulsion of some common everyday materials. This is caused by the buildup of free electrons on surfaces and may lead to electric sparks as the electrons jump from surface to surface to balance the charges (Wikipedia, Electrostatics). In ordinary materials the charges of electrons and protons are almost perfectly balanced and bound in atoms and molecules except in ions or where electric currents are present.

So the electrical force is simply the repulsion or attraction between the equal or opposite electrical charges of elementary particles. This is the force that binds particles together and creates all the matter in our universe. It's also the force that holds identical charges apart so matter doesn't all collapse in on itself. These binding energies give matter the precise structural balance it needs to create the specific atoms and molecules possible in our universe and all the chemistry and life that emerges from it.

So the electric force along with the strong force that overcomes it to bind positively charged protons together in nuclei are the two main forces that make atoms and molecules and everything made of them possible in our universe. The atoms that make up all the matter in the universe are electrical balances of nearly identical numbers of negatively charged electrons in orbitals surrounding an equal number of positively charged protons held together by the strong force in the nucleus.

Electricity, in the everyday sense, is not fundamental but emergent. Electricity is simply the *movement* of electrical charges through space. This can be in the form of loose electrons moving from atom to atom as a current through a wire, or as plasma where free electrically charged particles (ions) are moving through space as in lightning. The movement of loose electrons through a wire to form a current can be a continuous flow in one direction as in direct current or a back and forth flow of electrons in alternating current.

The electric energy that powers all the devices of our modern world is not the conversion of the intrinsic (internal) relative motion of electron charges but the energy of linear velocity of the electrons themselves moving through space through some electrical device. Electrons themselves are not used up from an electrical circuit as they power our appliances; that would involve the *mass* of the electrons being converted into a form of atomic energy and would produce an enormous unsustainable charge imbalance that would tear the atoms in the wire apart. The energy drawn from an electric current is some of the energy of their *flow* through the circuit. It's some of the relative motion of their flow that is converted into other forms of relative motion such as light, heat or mechanical motion.

And of course there must be an original external source of energy from a generator that is converted into driving the flow of the electrons through the power lines in the first place. Generators can be driven by the conversion of various forms of relative motion such as the energy of moving wind or water, the relative motion of heat from burning coal or nuclear fission, or the absorption of solar photons.

There is not nearly enough space here to describe all the common electromagnetic effects in terms of our theory. However note that when charged particles move their standing helical velocity density fields move with them as the change in motion propagates through the field at the speed of light. Changes in velocity (accelerations) of charges produce electromagnetic radiation because they manifest both changing electric and magnetic fields, which is what electromagnetic radiation consists of.

An electric current in a wire creates a corresponding circular magnetic field around the wire. Its direction (clockwise or counter-clockwise) depends on the direction of the current in the wire. This is the principle of the electric motor in which charges moving through a solenoid (current carrying wire coil) around a rotor magnetically rotates the rotor converting the relative motion of the electrons into the relative motion of the rotor.

The effect is reciprocal. A current is induced in a loop of wire when it's moved with respect to a magnetic field; the direction of current depends on that of the movement. Thus when a magnet is mechanically rotated with respect to a coil of electrical wire, or vice versa, its moving field of helical distortions acts on the wire in the coil to move its electrons along and generate a current. This is the principle behind electric generators.

There must be a closed (continuous loop) circuit for loose electrons to flow along a wire. Otherwise they would tend to pile up at one end and repulse each other there, and leave net positive charges at the other end that would tend to attract them back to that end. Thus electrons can't flow along a disconnected wire without being added at one end and discharged at the other.

The movement of electric charges either as the flow of loose electrons through a wire or as the movement of a magnet consisting of aligned spins, is basically two views of the same phenomenon. The movement of electric charges at one location induces the movement of electric charges in the other through their intermediary magnetic fields. If the charges are fixed in the material the material itself will move, while if the charges are loose they will move within the material. Both motors and generators operate as a result of the single principle that electric charges induce velocity vectors in adjacent electrical charges through intermediary magnetic fields causing them to move.

THE STRONG & WEAK FORCES

The strong force, or strong interaction, is the force that holds neutrons and positively charged protons together to form atomic nuclei. It's approximately 100 times stronger than the electromagnetic force, which it must be to overcome the mutual repulsion of the positive charges of the protons. This force is carried by gluons, which bind the quarks that comprise protons and neutrons. These quarks are bound by exchanges of gluons, analogous to how electrons are attracted to protons via exchanges of virtual photons, however there are 6 different 'color' charges for the strong force as opposed to the two plus and minus charges for electromagnetic force, and the single positive mass charge of the gravitational force.

The weak force or weak interaction is the last of the 4 fundamental forces of nature. The weak interaction is responsible for the radioactive decay of subatomic particles and it plays an essential role in nuclear fission. It is carried by the ±W, and Z bosons. It's of particular interest that the weak force is the only force that doesn't produce bound states.

The strong and weak forces and the electromagnetic force, are well understood in terms of the Standard Model, which predicts most but not all, of what occurs at the level of elementary particles. Notably it says nothing about dark matter or dark energy, and is unable to incorporate gravitation into the theory. Thus clearly a deeper more comprehensive theory is needed (Wikipedia, Standard Model).

Universal Reality suggests that progress will come from reinterpreting all these phenomena as computational processes which in concert generate the entanglement network science currently interprets as material structures in dimensional spacetime. We have suggested ways this might occur with respect to gravitation and electromagnetism and expect the strong and weak forces can be best explained in a similar manner.

Thus the strong and weak forces will likely be elemental computations on the data of elementary particles that conserve particle components so as to produce observational results in accordance with the standard model and whatever extensions are necessary to complete it.

Since all the forces of nature are energetic processes and Universal Reality proposes that all forms of energy are various forms of relative motion that dilate spacetime in a manner that transmits the relative motion to other particles in the field, it's reasonable to expect explanations of the strong and weak forces in terms of additional forms of velocity densities. In fact it's possible that the 4 forces somehow correspond to the different possible modes of vibration of space just as electron orbitals correspond to the possible standing harmonic waves in 3-dimensional space in atoms as explained in the section on Bound Particles in the upcoming chapter.

It should be noted that the science of the standard model is already taking small steps in this direction. It agrees in one important respect with Universal Reality's proposal that mass is actually a form of vibrational relative motion. Both protons and neutrons are composed of quarks but the masses of the constituent quarks account for only about 1% of the rest mass of the proton and neutron. Thus 99% of the mass of protons and neutrons is actually composed of the relative motion of the massless gluons that hold the quarks together (Wikipedia, Proton). Universal Reality naturally interprets this relative motion as an additional form of velocity density computed at the level of the entanglement network.

Here we have a clear case in which modern science has now discovered that mass actually is at least mostly relative motion as Universal Reality predicts. And since protons and neutrons make up almost all the mass of the visible matter in the universe, that means close to 99% of the mass of the visible universe is relative motion even according to contemporary scientific theory.

COMPUTING QUANTUM REALITY

EVENTS & THE ENTANGLEMENT NETWORK

At its most elemental level the universe is computed in terms of particle events. At every P-time tick the data state of every particle in the universe is simultaneously recomputed. When particles interact including when particles interact with fields, including the field of empty space, each *coherent process* is computed as an individual process by a separate *application* of the universal processor. This includes the computations of bound particles in atoms and molecules. In this way all data states of all particle interactions are recomputed at every P-time tick.

Every event conserves the total amounts of all properly defined particle components of all particles. This means that when events transform particles into other particles the conserved amounts of all particle components are exactly redistributed among the emitted particles.

Particle component conservation *entangles* all emitted particles on each of their particle components separately because each is separately conserved. Entanglement simply means that the values of all particle components of emitted particles bear fixed relationships to each other. This relationship is numerically exact when only two particles are emitted or relational when multiple particles are involved. It's important to understand that entanglements are not isolated occurrences in scientific laboratories as some might assume but relationships among particles generated by *every particle interaction* in the universe.

Thus the entire universe consists of an *entanglement network* of entangled particles because all particles are part of a universal network of chains of particle events. All particles in the universe have interaction connections with all other particles going back to the original event of the big bang. Thus the observable universe itself consists of the entanglement network of all its actual particles, or more precisely of the entanglement network of all its particle components. This entanglement network contains all the information of the observable universe including the data of all particle structures and their dimensional relationships. Observers interpret the entanglement network as a material universe in a pre-existing physical spacetime in their simulations of reality even though it actually consists entirely of the data of particle component relationships and is constructed on the fly by particle events.

For example if a total mass-energy of e is conserved between only two emitted particles then knowing the mass-energy of e_1 allows us to know the mass-energy of the other exactly since $e_2 = e - e_1$. However if there are 3 or more particles emitted the relationship between any two particles is no longer exact since $e_2 = e - (e_1 + e_3 \ldots)$ and the individual mass-energies of e_1, e_3, etc. could vary and still satisfy the equivalence. However the possible variation is highly constrained by the equivalence so there still is considerable entanglement. This is true of the entanglement of each type of particle component separately since each type is conserved and entangled separately.

Because each type of particle component is conserved separately the entanglement network consists of separate layers for each type of particle component. The separate layers are joined together at nodes representing particle interactions or events, and the multiple layers connecting nodes represent particle components packaged as individual particles. All the layers of each line connecting event nodes taken together represent a particle since each particle is composed of its particle components.

Entanglement is easy to understand, as it also occurs at the classical level. If we take a cookie and break it in two, the two halves of the cookie are entangled on their masses because the sum of their masses equals the mass of the original unbroken cookie. If we measure the mass of one half we immediately know what the mass of the other half is. So the basic concept of entanglement is quite simple. *Every* conserved quantity is automatically entangled when divided into multiple parts or reapportioned among different entities. Entanglement due to conservation is a universal law.

All the components that make up particles are conserved and entangled by particle events. These include particle identity (lepton and baryon number), mass-energy, the other charges, spin, weak isospin, and space and time parity (handedness in space and time). Particle components are the little bits of what it takes to make something real and actual in our universe because particle components make up the elementary particles that compose all structures.

Particle components are numeric data entities rather than physical entities. They, rather than particles, are the actual fundamental elements of the observable universe because it is they rather than particles that are conserved through all particle events. Particles are composed of pre-defined valid sets of particle components stored as data templates in the complete fine-tuning.

Momentum is also conserved within the limits of the Uncertainty Principle but importantly the positions and velocities of particles that are used to dimensionally express the conservation of mass-energy and momentum are not conserved. Total mass-energy is conserved due to the equivalence of all forms of mass and energy as different forms of relative motion, of spatial velocity.

Successive particle events entangle more and more particles in entanglement chains. And since particle events typically involve multiple interacting and emitted particles each of which may interact in subsequent events all particle events taken together create a vast entanglement network that includes all particles in the universe and stretches via common entanglement links back to the original event of the big bang.

The *current* present moment surface of the entanglement network is the entire actual universe in the current present moment. The current data state of the entanglement network is the actual universe, which is a logico-mathematical data structure. It's not a physical universe but it contains the data we humans interpret as a physical universe.

The entanglement network is a data structure that is computed by the fundamental program as it simultaneously computes all individual particle events. It's the data of the relationships among all particles in the universe produced by elemental events and thus it includes both the mass-energy structure and the spacetime dimensionality of the universe in a single unified structure. It encodes the logico-mathematical structures of both mass-energy and spacetime in a single universal information structure. In this way spacetime and mass-energy structures are computed together by quantum events. Spacetime is not a preexisting container in which events occur or are even computed; it's the consistent set of all dimensional relationships among particles computed by events.

THE SOURCE OF QUANTUM INDETERMINACY

A major unanswered problem of physics is why the universe, in particular its dimensionality, is 'fuzzy' or indeterminate and random at the quantum level. Universal Reality provides a simple and convincing answer in terms of the same space versus time allocation of processor cycles that computes a relativistic universe at the classical level.

Universal Reality proposes that all forms of quantum indeterminacy are the result of random space versus time *oscillations* in the processor that computes all the events in the universe. Specifically the zero-point energy fluctuations of the quantum vacuum, the related Uncertainty Principle, and the wavefunction representations of particles are all manifestations of very fine scale oscillations between processor cycles allocated to computing velocity in space and velocity in time.

As the processor computes the space versus time evolution of processes in accordance with relativity as explained previously there is a simultaneous fine scale oscillation of these cycle allocations. This manifests as an indeterminacy of space versus time at the quantum level. In effect the reference background against which observable dimensional spacetime values are being computed is itself continually oscillating between space and time and this results in the observed spacetime 'fuzziness' of quantum phenomenon. This makes sense because all such fuzziness reduces to indeterminacies of space versus time upon examination.

This is demonstrated most clearly in the Uncertainty Principle (Wikipedia, Uncertainty principle). The uncertainty of energy, which is spatial velocity, versus time, and that of momentum (again spatial velocity) versus position are clear examples of a source in a minimal but fundamental conflation of space and time. If observable values are always determined with respect to a dimensional background that is itself oscillating between space and time at the quantum scale the observed values will inevitably be uncertain in space and time around the value of the oscillations.

So the allocation of processor cycles between computing space versus time velocities provides a very simple and convincing mechanism to explain both quantum fuzziness and general relativity at the classical scale. Both quantum and relativistic processes become part of the same computational system.

WAVEFUNCTIONS & DECOHERENCE EVENTS

The entanglement network consists of event nodes connected by particle that link them. At Level 0 events are exact calculations of the conservation of individual particle components. However conserved energy values must be translated into dimensional positions and velocities to observably manifest. Whenever dimensional values of space and time are computed the oscillation of processor cycles results in uncertainties of space versus time variables and these space versus time uncertainties take forms describable by wavefunctions.

Due to the processor oscillations the computation of space versus time is inherently uncertain at the quantum level even though there is an underlying Level 0 numeric exactitude demonstrated by the exact trajectory of the wavefunction itself and its exact deterministic time evolution as described by the Schrödinger equation (Wikipedia, Schrödinger equation).

The exact trajectory of the wavefunction itself reveals the exact underlying Level 0 numeric data entity trajectory of the particle. However the form of the wavefunction as it travels represents the probabilistic mapping of that entity into observable dimensional values. So at every P-time tick the exactly conserved energy values computed by events are expressed as positions and velocities whose space versus time values are conflated at the quantum scale by the random oscillations of processor cycles between computing velocities in space and velocities in time.

The computational evolution of the particle's exact trajectory into observable dimensional space and time values derives from the tick-by-tick oscillations of the space versus time processor oscillations. These oscillations occur randomly at each P-time tick for each processor application computing a separate coherent process but the form of the oscillations over time is that of the probability amplitudes that wavefunctions describe.

Thus wavefunctions and the Schrödinger equation correctly describe the time evolution of particles but what they are actually describing is the time evolution of the allocation of the processor cycles that compute that evolution. Wavefunctions don't describe the fuzzy form of particles with respect to an exact pre-existing spacetime but the actual computation of dimensional spacetime from particle events. It's the spacetime itself that is intrinsically fuzzy at the quantum scale as it's computed because the processor cycles that compute dimensionality continually oscillate back and forth between computing space versus time in a random manner at the quantum level.

Thus Universal Reality provides a huge paradigm shift in the understanding of quantum processes. They are not composed of particles that are intrinsically fuzzy with respect to an exact universal pre-existing spacetime but spacetime itself in the process of being created by the conservation of mass-energy by quantum events that is created as dimensional fragments that are intrinsically fuzzy with respect to each other.

Every entanglement event creates dimensional relationships among its emitted particles to conserve their total mass-energies. The dimensional relationships among all the particles emitted by an event form a *coherent dimensional fragment* that continues to be computed as a single process by dedicated applications of the universal processor. Because the time evolution of all the particles emitted by an event are computed as parts of a single process the entire process remains *coherent* and all the entanglement relationships are preserved.

Thus there is an on going computational connection between all entangled particles and this is the secret to the apparent spin entanglement paradox. Entangled particles are computed as aspects of a single process and this is how a measurement of the spin of one of those particles immediately affects the spin of the others. This is explained in more detail shortly.

The exact numeric dimensional values of particles at Level 0 is not observable as previously explained. The totality of all such values constitutes the absolute computational space with respect to which all linear and rotational relativistic motion is relative to.

It's the allocation of processor cycles to the time evolution of exact particle trajectories that computes the wavelike dimensionality of particles. Subsequently *observable dimensionality* consists of particle measurements including measurements in the generic sense of particles measuring each other's dimensionality in mutual interaction events. All our human scientific and sensory observations are ultimately scaled up dimensional values produced by individual particle-particle interactions in which particles measure each other's dimensionality by interacting.

What humans interpret as a continuous physical spacetime is actually a reification of the logico-mathematical consistency of all dimensional measurements in which they participate. There simply is no actual empty physical spacetime container in which events occur. It's a convenient illusion produced by our simulations based on the internal consistency of individual particle measurements. Our minds pin the dimensionality of individual observations together and extrapolate an imaginary spacetime container on that basis with us at the center.

Getting back to the entanglement network upon which this is all based, events conserve all particle components including total mass-energy. For particle components to be conserved they must have exact values. This is not a problem because the processor cycles are always computing exact dimensional values in every P-time tick even though those values vary probabilistically from tick to tick.

Assume that an event occurs when two particles from two separate processes both take on the same random dimensional values. This is a simplification as events can occur when colocations are not exact but the process is essentially the same. Because all processes are being computed simultaneously in the universal processor the processor immediately 'knows' when particles collocate.

The type of event is determined by the energy, degree of colocation, and types of particles involved. There are two general types of events, scattering events in which the particles themselves acquire new trajectories, and particle events in which the particles break apart into their particle components to form new particles with a new distribution of those same particle components.

Events cannot occur unless the total amounts of all particle components can be validly allocated among emitted particles. However when alternate sets of particles could satisfy this requirement all those alternate sets are a possible outcome. This is referred to as the 'Totalitarian Principle' (Wikipedia, Totalitarian principle). In this case nature decides randomly which set to produce based on their probabilities, which generally has to do with the most efficient (entropic) distribution of energy.

This is demonstrated in the very high-energy collisions of particle colliders such as the LHC where scores of new particles may be produced by a collision of only two hadrons. Many different sets of particles could be produced so long as all particle component totals are conserved.

With high enough energy new particles are literally created out of the vacuum to conserve the incoming energy. For this to occur particle components must be actualized in particle-antiparticle pairs out of the vacuum. In this case the total particle component values (other than mass-energy) of the newly created particles sum to zero so the original totals are conserved.

When an event occurs a new process begins that is now computed by a new application of the processor. In general the entanglements the particles had with the other particles of their previous particles are instantly resolved as the particles are separated from their previous processes to become parts of the new process that computes the new event.

Thus the new process begins with the dimensional values of the event and these values are then converted into the total mass-energy of the process so it can be conserved and redistributed among the dimensional variables of the particles emitted by the event.

This 'freezing' of the random dimensional values of two interacting particles by an interaction is called decoherence (Wikipedia, Quantum decoherence). It is often incorrectly interpreted as the instantaneous 'collapse' of wavefunctions by a measurement.

Once an event occurs the process begin all over again. All particle components of the event including the total mass-energy are conserved and redistributed among the emitted particles, which continue to be computed as aspects of a single new coherent process. The conservation entangles each of the particle components among all the emitted particles and the coherent oscillations of processor cycles in the process continue to exactly relate the dimensional and other particle component values of all particles in the individual process. The result is an evolving dimensional fragment that carries the coherent dimensionality of the entangled particles in the process.

Because the dimensionality of all the particles in a coherent process are computed by a single application of the processor the random values the processor produces at every P-time tick are all interrelated so as to preserve all the conserved values of the originating event. This is how the processor keeps track of the entanglement relationships among all the particles of each coherent process. It's because they are all being computed as parts of a single process no matter how they may be separated in observable spacetime. Nonlocality is no problem at all in a computational universe because everything always exists together in the non-dimensional computational space in which it's computed.

The processor that computes all processes automatically computes probabilistic dimensional values for all particles at every P-time tick. This is required so particles can be tested for colocation because otherwise the processor wouldn't know when or where to initiate events.

These random dimensional values computed at every tick are in effect *virtual decoherences*. They are exact dimensional values produced randomly at every P-time tick by the processor cycle oscillations for all particles. They are produced randomly on the basis of the probability distributions of how the oscillations evolve from tick to tick which again is accurately described but misinterpreted by the wavefunction model.

In the case of particles the magnitude of dimensional fuzziness is not only a function of the oscillations but also of the individual particle components. This is how different particles in the same dimensional fragment can have the varying dimensionalities their different wavefunctions describe. The dimensional fuzziness itself is always due to the processor oscillations but its magnitude is affected by the particle components for example the masses of the particles.

So there is an exact numeric dimensional background that consists only of exact particles with exact dimensional values. This is the dimensionality of the computational space in which everything is computed and it consists only of data and it is completely unobservable except by its effect on observable dimensionality.

Observable dimensionality is computed probabilistically with respect to this exact background due to the random processor oscillations between space and time cycles. This results in observable dimensionality being fuzzy between space and time at the quantum level and this manifests as the well-known dimensional fuzziness of quantum processes in wavefunctions, the Uncertainty Principle, and the related zero-point energy fluctuations.

Each separate application of the processor to an individual dimensional fragment is characterized by its own probability distribution of possible randomness over successive P-time ticks. This is accurately described by the quantum wavefunction model, however Universal Reality interprets wavefunctions as descriptions of the time evolution of processor cycle oscillations rather than particles smeared out in a fixed continuous pre-existing spacetime because such a pre-existing spacetime simply doesn't exist.

Thus the entanglement network that encodes the data of the entire observable universe consists of exact (within uncertainty limits) event nodes connected by particles whose observable dimensionality is probabilistic due to the fine space versus time oscillations of the processor cycles that compute it. The universe consists of a vast dynamic entanglement network of more or less exact particle events linked by dimensionally probabilistic particles whose probability distributions move along exact trajectories.

Due to the non-proportionality of particle masses, a portion of conserved event energy must be expressed in terms of velocities and positions, so an observable dimensionality is created in which velocities and positions can manifest the excess energy as the kinetic energy of motion. This is how and why quantum events create spacetime, as the only available way for events to exactly conserve mass-energy. It's the non-proportionality of particle masses that requires the creation of a dimensional spacetime in which mass-energy can be conserved in terms of velocities and positions.

The result of this process is the entanglement network whose current P-time surface is the entire actual universe in the current present moment. The observable universe of science and the senses consists entirely of observable values ultimately produced by particles measuring particles in decoherence events.

The observable universe is a consistent logico-mathematical data structure in that the relationships among all observables are internally consistent according to the laws of physics. Even though particle dimensionality evolves probabilistically between events it does so in a logico-mathematically consistent manner. And at classical scales the covariance of general relativity is part of this consistency.

QUANTUM GRAVITY

A major problem in modern physics is the apparent incompatibility between quantum theory and general relativity. This is due primarily to the inconsistent manner in which the two theories view spacetime. Quantum theory considers spacetime an exact static preexisting container within which quantum events occur, while the preexisting spacetime of general relativity is dynamic and shaped by the presence of matter.

Universal Reality proposes a completely new approach in which there is no preexisting spacetime and what we call spacetime is the consistency of dimensional relationships produced by quantum entanglement events. In this theory spacetime is computed by quantum events as the dimensional relationships among mass-energy structures as they evolve.

To incorporate relativity into this model quantum events simply scale the dimensionality they compute proportional to the mass-energies of the particles involved in the event and this automatically produces the dynamic curved spacetime of general relativity or in our model fields of intrinsic vibrational density. Every quantum event simply computes the masses of the particles involved as fields of vibrational density in the surrounding fabric of space as it's constructed. In this manner the myriads of ongoing quantum events of all the particles of say a planet are additively computed as a gravitational field consisting of mass vibrations in the space surrounding the planet and as we have seen this produces the relativistic effects of a gravitational field.

The relativistic effect is extremely weak at the quantum level due to the relative weakness of the gravitational force so the standard Schrödinger equation accurately describes events at the quantum level without it, but for large aggregates of particles the simple scaling of dimensional relationships by the presence of mass-energy produces the spacetime of general relativity in a rather straightforward manner as fields of velocity density. Once the paradigm of quantum events creating spacetime in the form of an entanglement network is accepted this practically suggests itself. Observable spacetime is then the interpretation of the dimensional aspects of the entanglement network in the simulations of human observers.

In relativistic terms this is how mass tells space how to curve. The reciprocal effect, how space tells mass how to move, is due to the allocation of fewer processor cycles to computing velocity in time after some of the fixed number of cycles are used to compute spatial velocity. And the total spatial velocity is simply the sum of linear velocity and the velocity density of gravitational fields as previously explained.

So the presence of mass (energy) in particle interactions scales the dimensionality created by quantum events by adding velocity density fields, and the intrinsic velocity of the fields in turn scale the allocation of processor cycles to compute the time dilations of events in the field. This gives us a neat approach to explain how quantum events create general relativistic spacetime by scaling the dimensionality of mass-energy structures as they compute them.

A VISUAL MODEL OF THE UNIVERSE

Universal Reality proposes the computational mechanics of the processor that computes the universe is the source of both general relativity and quantum phenomena. There is a very useful visual model that makes this clear when correctly interpreted.

Imagine the universe with one spatial dimension suppressed for clarity as a 2-dimensional surface. This surface represents a section of the 3-dimensional space of the universe in the current universal present moment, the only moment that actually exists. At every P-time tick the current state of the surface is completely recomputed to produce a new state of the entire surface. The entire evolutionary history of the universe corresponds to the succession of all past surfaces of which only the top one is the real and actual universe in the current present moment.

If we wish we can visualize the entire recomputed surface rising with each P-time tick so the entire stack of surfaces represents the evolution of all the data of the universe through time. However only the topmost surface corresponds to the entire actual universe in the universal current present moment.

Now to be clear the universe is not actually the physical dimensional structure this visualization implies. It's a graphical visualization of what is actually entirely a data structure. It's a model of the observable dimensionality of the data of the universe in exactly the same sense that dimensional data in a computer program consists only of numbers in memory locations but can be graphically displayed as objects in space on a computer screen. The same is true of the actual universe which consists of logico-mathematical data in the form of the entanglement network in computational space but which observer minds visualize graphically as a physical spacetime universe.

Note also that even observable spacetime consists only of individual observable dimensional values rather than a continuous encompassing physical space. So the surface model is misleading, as observable dimensional values are actually individual data points that a surface could be extrapolated from as it is in our mental simulations of the universe. It just so happens those points always fall where a surface would be if it existed. Our minds note the consistency of where numeric valued points fall and extrapolates that into a continuous empty space, but this empty space is a fiction of our simulation of reality rather than part of reality itself.

The data of a computer model exists as numeric data in the computational space defined by the program. Similarly the data of the universe exists as data in the computational space of the quantum vacuum, which is the medium or substrate of existence and thereby gives all data within it actual existence as a real actual universe.

A major difference between the two types of models is that the quantum vacuum in which all the data of the observable universe exists acts as a single universal processor that computes all the data of the universe simultaneously, whereas in computers the processor(s) computes data stored outside the processor sequentially fetching single units of code and data into the processor to compute them. In the quantum vacuum all the data of the universe always exists within the universal processor that computes it. Only in this manner can internal consistency among all ongoing processes be guaranteed.

So our visual model of the universe is just a graphical visualization of how spacetime could be interpolated from observable dimensional values as they are computed from elemental data. However the model is extremely useful in understanding how the universe is computed if we keep these caveats in mind.

Each successive surface layer of the universe is computed from the data state of the previous layer. This takes place by separate applications of the universal processor acting on the data states of all the individual coherent processes in the universe. Thus each individual process in our model is computed by an individual *application* of the universal processor. In this manner every coherent process in the surface is in a continual state of recomputation.

Now imagine each individual computational process of the universe occurring within the surface as it progressively rises with each P-time tick. The surface is not a pre-existing physical spacetime but represents the unobservable background reference space in terms of which observable dimensional values are computed. Rather than being a dimensional space itself it's the standard reference with respect to which dimensionality is computed.

Now imagine each point in the surface is finely oscillating back and forth sideways (in space) and up and down (in time). Thus the dimensional values of the process being mapped to the background reference become intrinsically uncertain at the scale of the oscillations. The process itself (the particles involved) is exact but its dimensionality is inherently fuzzy because the reference space and time rulers that measure it are continually oscillating.

Now visualize each individual computational process as a separate plane because each process is computed by a separate application of the processor, by a separate call on the elemental program. Thus each individual process will be subject to a different processor oscillation sequence from tick to tick as it's being computed. All aspects of each individual process are computed as part of a coherent whole so that each part has an exact relationship to all other parts in this respect.

However because each process is an individual application of the elemental program computed by its own application of processor cycles, every individual process and dimensional fragment can be visualized with respect to a separately oscillating reference background with respect to which its dimensionality is being computed. So there is a different oscillation pattern for each coherent dimensional process but the dimensionality that's computed for separate particles is also a function of the details of each individual particle of the process.

Each dimensional process computes what can be considered a separate dimensional fragment because its dimensionality is inherently fuzzy with respect to all others.

We can retrieve the overall view of the model by collapsing all the individual process planes of each P-time surface back into the surface so long as we remember that each individual plane corresponding to an individual coherent process and dimensional fragment is actually oscillating differently because it's being computed by a separate application of the processor with it own individual oscillatory pattern.

The result is that entangled particles are still well described by their individual wavefunction equations. But we also have a non-paradoxical explanation for entanglement and quantum indeterminacy in general in our model that is lacking in current quantum theory. And because dimensionality is computed along with particle structures our model is also consistent with general relativity as well as quantum theory.

Though the dimensionality of all individual processes oscillates differently at the quantum scale the *classical scale* dimensionality of each process's background surface appears identical because the oscillations are not detectable at the classical level and so the classical universe behaves in a relativistic rather than quantum manner.

Thus Universal Reality neatly unifies quantum theory and general relativity in a single model by demonstrating how they are both computed by the same space versus time velocity allocation of the processor cycles that compute the mass-energy and dimensional structures of the universe in the form of an entanglement network of the particle components that make up elementary particles.

RESOLVING THE SPIN ENTANGLEMENT PARADOX

The processor oscillation model explains the coherence of entangled particles in a dimensional fragment. They were created together as part of a single coherent process and continue to be computed together as part of that single process so long as they remain entangled.

For example spin entangled particles initially created with equal and opposite spins in a single process of spin conservation, a single application of the elemental program, continue to move apart as part of a single coherent process that computes the evolution of both particles with the same processor cycle oscillations at each P-time tick.

The spin entanglement paradox is of course considered a major unsolved problem of quantum theory, which assumes quantum events occur separated in a physical spacetime. But in a computational model everything is *automatically local* because everything is computed in a single universal computational space with no physical expanse so computing non-locality and spin orientation entanglement is no problem at all so long as a single coherent process computes it.

The apparent paradoxes of quantum reality all turn out to be with respect to a mistakenly assumed pre-existing spacetime contained that doesn't actually exist. By deprecating this pre-existing spacetime container and showing how dimensionality is actually computed by quantum events all quantum paradoxes are resolved. The same model simultaneously unifies quantum reality and general relativity.

In our computational universe all quantum paradox is resolved because there is no actual spatial distance between the data of separate particles. Everything (all the data of the universe) is always in the same place (the single processor) at the same P-time. Thus the entangled values (for example spin orientations) of two particles at different locations are automatically related because they are computed as a single coherent process with the same pattern of processor oscillations.

In a computational universe there is no problem of non-locality or faster than light transmission of information because everything is always in the single processor that computes the entire universe at every moment. This straightforward non-paradoxical explanation of spin entanglement is excellent evidence in support of our theory.

So there is an inherent computational connection between the spins of spin entangled particles, as there very obviously must be. This occurs automatically because they are computationally related by the conservation of the event that created them. Since they are computationally connected and conserved then it's easy to see why decoherence of either one to an exact orientation automatically decoheres that of the other to the exact opposite orientation because only this preserves the original conservation of spin.

Thus if the spin orientation of either particle is measured the process that computes the measured orientation is already simultaneously computing the spin orientation of the other particle and when the orientation of the other is fixed so is that of its partner since they were always being computed together at every step of the way since their mutual creation. The spin entanglement paradox is no longer paradoxical but the straightforward consequence of a computational universe in which entangled particles are computed together as aspects of a single coherent process as a separate coherent dimensional fragment.

In effect the decoherence of the spin orientation of either particle via a measurement automatically decoheres the spin orientation of the other. This is a general principle in Universal Reality; that a decoherence of any of the entangled particles in a dimensional fragment is automatically accompanied by exact values for the others because they are all being computed as aspects of a single process by a single application of the elemental program and processor.

If there are only two particles the decoherence is complete but if there are multiple other entangled particles in the dimensional fragment their entanglements are adjusted by the decoherence of the measured particle which drops out of its dimensional fragment to become part of the new dimensional fragment computed by the process computing the new event.

To be more accurate all the particles in a single dimensional fragment *virtually decohere* to exact dimensional values at every P-time tick. This means their virtual decoherence values are all interrelated so as to preserve the conservation of the event that created them. Thus the positions and velocities of all particles in the dimensional fragment are exactly related at every P-time tick.

However if one particle's virtual decoherence is converted into an actual decoherence by a colocation interaction with a new particle the dimensional fuzziness of the other particles in the dimensional fragment then continue on as before adjusted by the fact of the actually decohered particle dropping out to join the newly created dimensional fragment of the event that decohered it. However in the case of entangled spin orientations of only two particles the actual decoherence of the spin orientation of one particle produces a permanent decoherence of the spin orientation of the other particle in its dimensional fragment.

This solves the problem of why spin orientations are always opposite, but why do they both remain fixed once one is measured? This is simply because spin orientation is not intrinsically affected by processor oscillations because it's an orientation in space that has no time component unlike a velocity. Therefore the oscillations of temporal position and velocity don't affect it and neither do oscillations of spatial position and velocity.

But why aren't spin orientations determined when they are created? Why are they created equal and opposite but with no specified orientation to the background? This is because total spin would be conserved by any orientation of the paired particles so long as they were equal and opposite. Thus there is no computational rule to determine any specific paired orientation and whenever there is no computational rule either a choice is immediately made randomly if it can be or is deferred until specific rules apply.

In the creation of paired spins there are no determining rules and the choice is deferred until it is made in a subsequent event that has rules that force the decoherence of some actual spin orientation. The new event must decohere the spin to an exact value so as to be able to conserve the total spin of the new event. This is an important general principal that underlies all quantum randomness.

FIELDS & VIRTUAL DECOHERENCE

All particles must *virtually decohere* to exact values (within uncertainty) at every P-time tick. This enables the elemental program to test for any colocation that would trigger an actual particle event. Colocation means the particles have close numeric dimensional positions, not that they are actually close in a physical space.

If two particles are collocated an event is triggered and an actual mutual decoherence occurs with the values of the virtual one. An actual decoherence to exact dimensional values is necessary because events can only conserve exact particle component values. Thus events always require actual decoherences. The virtual decoherence is actualized in the event and the exact decohered dimensional values of all particles are then conserved in the event which begins to be computed as a new process by a new application of the processor and elemental program.

When a decoherence event occurs new trajectories for all resulting particles are created originating from their decohered values and begin to be computed together as coherent aspects of a single new process in a new dimensional fragment.

This model covers particle interactions but it's not complete. Particles interact with fields as well as other particles, so fields must be added to our decoherence model. In particular we need to include the interaction of particles with electromagnetic and gravitational fields in a manner consistent with general relativity.

Fields are a mechanism by which particles interact with other particles at a distance. Charged particles attract or repel each other at a distance and the mass charge of gravitation interacts with all other particles at a distance through gravitational fields. Fields are intrinsic aspects of charges. The field is an actual part of the charge.

To include the effects of fields charged particles are modeled as spherical fields of velocity density rather than points. As discussed earlier the charges of the four forces are all forms of mass-energy and thus different forms of relative motion whose velocity densities are their strengths. An alternate but largely equivalent model is to treat fields as exchanges of particles, which may be more useful for the strong and weak forces, which are more clearly mediated by their respective bosons. Recall that charges including their fields and space itself are both two aspects of the same phenomenon of spatial velocity.

Thus fields are dimensionalized as spherical clouds of intrinsic relative motion surrounding charged particles and traveling with them as they move. Everywhere a particle goes its field accompanies it because the field is an actual part of the particle. The charges of particles actually are intrinsic velocity density fields centered on their nominal particles. Charges are not points but velocity density fields with forms corresponding to the type of charge. These velocity density fields are equivalent to the dilation or curvature of spacetime in the general relativity model.

A particle located in a field experiences its intrinsic velocity and thus acquires an additional spatial velocity that reduces its temporal velocity in accordance with the STc Principle. Each point in the field can be characterized by an STc tilt that slows its clock proportional to its velocity in space so its total velocity through space and time is always equal to c. This produces the correct time dilation and other effects of general relativity.

Fields are part of charges so their dimensionalities oscillate with the dimensional oscillation of their central particle. The dimensionality of the whole field oscillates with its central particle at each P-time tick because it's computed relative to its center.

Particles interact with fields; so entangled particles computed as part of the same coherent process can be affected differently by fields because one particle can be deeper into the field than another. Thus the probability distributions of their dimensionality with respect to the field can be different even though they are parts of the same dimensional fragment and being computed as part of a single process.

In order to compute the movement of particles through fields including the 'null' field of 'empty' space virtual decoherence events must take place at every processor cycle. Every processor cycle virtually recomputes the dimensionality of all particles in the observable universe as aspects of a single universal event. All particles virtually decohere and are tested for field interactions and colocations with other particles that might trigger actual particle events.

At every processor cycle every wavefunction and dimensional fragment, including their fields, is tested against others to determine the probability of interaction and events are computed on this basis. If particles interact they decohere as before, but if a particle's virtual location intersects that of a field, then the effect on the particle is calculated and the trajectory of the particle is adjusted accordingly.

Since field effects are energetic they modify the trajectories of affected particles. The total conserved relative motion of the dimensional fragment is adjusted by the amount of the effect on the individual particle. There is an instantaneous virtual decoherence that produces an exact random position and velocity of the particle relative to the field so the effect can be computed. Then the total conserved relative motion of the dimensional fragment is adjusted by that amount.

The entire process repeats with every P-time tick for all particles. There is a continual series of virtual decoherences at every P-time tick of all the entangled particles of all dimensional fragments that modifies them by the field interactions of their individual particles. There is still an entanglement carried by the dimensional fragment, but it now carries the modified total mass-energy rather than just that of the event that created it. It's the original relative motion adjusted by the cumulative effects of field interactions on its member particles at any point in time.

To what extent are entanglements maintained when particles traverse fields of different strengths? Are they still being computed by a single process and its specific oscillations or are they now affected by the oscillations of the process computing the field? This apparently depends on the strength of the particle-field interaction.

The passage of a wavefunction through the null field of empty space obviously doesn't cause it to lose coherence or all particles would exist in a decohered state all the time, but there must be some process of virtual decoherence at every tick in case particles interact strongly enough to produce an actual decoherence.

A particle passing through a field is affected by the field but the passage doesn't affect the field. Of course if the particle carries its own field the fields will additively interact and each particle will affect the other but the mere fact of a particle passing through a field doesn't affect the field. There is no influence back to the central charge creating the field other than via the field of the passing particle. This is key because it means there can be no actualized mutual decoherence from field interactions alone. There is only an effect on the passing particle, not on the particle generating the field.

One could argue that if fields are relative motion then their effect on a particle is energetic and should conserve energy by affecting the field but we can replace the fields as relative motion model with the equivalent model of fields as STc tilts or curved spacetime and the effect is now recognized as inertial so this is not really a problem.

Thus particles in fields don't actually decohere with the fields. There is an instantaneous virtual decoherence as there is with everything at every P-time tick so that the instantaneous effect of the field on the particle can be computed but there is no actual decoherence of the particle and thus no mutual decoherence of it and its entangled particles in their dimensional fragment. However the effect on particles in fields does progressively affect their coupling with entangled particles and gradually weakens the entanglement.

This model enables particles to pass through fields that change their trajectories as well as null fields that maintain their trajectories without actually decohering them or their entangled particles, which is what we actually observe.

So actual versus virtual decoherence depends on the strength of the interaction a particle has with its field environment. In general interactions with fields are not sufficiently strong to decohere particles though cases of scattering in which two particle trajectories are changed by each other's fields are probably best modeled as mutual decoherences. One needs to test the effect in particular circumstances to determine this. Our model covers both cases.

Another issue is that the quantum vacuum background is itself full of zero-point energy and its basic dimensional constituents fluctuate from tick to tick. This induces an inherent fuzziness of all dimensional values from tick to tick at the elemental level, which at the classical aggregate level averages out. This produces some level of fuzziness in the mutual positions and velocities of all particles though apparently not spin orientations.

This aspect of the model also explains the partial decoherences observed in some situations and those instances in which particles can be gently measured without decohering. This effect should be a measurable test of the model.

DIMENSIONAL FRAGMENTS

In our model each individual coherent process is represented against its own oscillating plane of incipient dimensionality, each with its individual spacetime fuzziness at the quantum scale. However at the classical scale the overall forms of all planes merge to the local relativistic spacetime velocity density field. This makes it possible to computationally test for particle colocations as all particles virtually decohere with every tick.

When particles collocate in computational space particle events occur and begin to be computed as a single new process. The particles are moved from their previous processes to the new process and their entanglement with their previous partners is broken and new entanglements are produced by the conservation of the particle components of the newly interacting particles.

Thus in the universe at large as particles continually interact their dimensionalities continually merge and are realigned, and in turn their relational connection with previously entangled particles is progressively diluted and previous dimensional fragments are gradually reduced as particles leave previous dimensional fragments and become part of new ones.

In this manner the entanglement network of all particles is continually shuffled and updated. At any point in P-time the current entanglement relationships among all particles is the data of the entire universe, both in its particle structure and dimensionality.

Each oscillating plane of a separate coherent process can be considered a dimensional fragment, and observable dimensional values are produced when dimensional fragment planes transiently align as their individual particles decohere with those of other planes. In this manner the dimensionality of the observable universe is continually rebuilt from the observable actual decoherences of particles from different dimensional fragments.

If we now take an orthogonal (non perspective) view from the top and collapse all the planes of separate processes back into a single surface of the present moment of the observable universe we arrive at a picture that is superficially much like a standard interpretation of quantum theory. The observable universe now consists of the current set of all decohered dimensional values, which together form a consistent logico-mathematical entanglement network in which the decoherence events producing the observable values are the nodes.

The event nodes of this network are connected by links, which are dimensionally fuzzy and could well be interpreted as particle wavefunctions. Thus we have a theory that looks like quantum theory on the surface, but which is fundamentally richer and consistent both with general relativity and a non-paradoxical view of quantum events.

This visual model is not quite complete and needs to be tweaked because entanglement is not always either or, and the separate planes (dimensional fragments) making up the surface may not always be entirely independent. In general events dilute but don't completely destroy entanglements with previous partner(s) in the dimensional fragment. There always remains some relationship among all particle component values in the entire entanglement network including those of mass-energy across all particles. This underlying logico-numeric consistency among all components of the entanglement network is in fact the consistent structure of computational space with respect to which all observational dimensionality is ultimately relative to.

However the *coherence* of entanglements is lost in decoherences because the same single process as before is no longer computing the entangled particles. The decohered particle is now being computed by the same coherent process as its newly entangled partner(s) rather than that of its previous partner(s). Only a single application of the processor with its unique oscillation pattern can compute any one process. Thus if the newly entangled particle is now being computed by a new process it loses coherence with the previous process that is no longer computing it.

Thus there remains an underlying numeric relationship among the dimensional values of all particles in the entanglement network even though only the currently entangled sets are being computed as separate coherent processes with a single application of the processor with its unique processor cycle oscillation pattern.

Thus all particles in the entanglement network will have consistent numeric relationships (within Uncertainty limits) among all their particle component values. They will not always be exactly one-to-one but they will never be inconsistent. And the set of all decohered dimensional values, which is the set of all *observable* dimensional relationships among all mass-energy structures, is the set of all dimensional values observers record and then interpret as a physical spacetime.

Thus observable spacetime consists of interrelated dimensional values (positions and velocities) sewn together by all decoherences from the dimensional fragments created by quantum events conserving and entangling the mass-energies of emitted particles.

Observable spacetime forms a consistent logico-mathematical structure because events exactly conserve mass-energy. While mass-energy is probabilistically manifested as dimensional observables at the quantum scale observable dimensionality seems exact at the classical relativistic scale. It's the logico-mathematical consistency of observable dimensional values that humans physicalize and interpret as an encompassing spacetime container within which events occur.

From an overall view particles and events consist of exactly conserved numeric particle component values computed by individual applications of the elemental program to each separate coherent process. Because of the non-proportionality of particle rest masses conservation requires a dimensional spacetime in which positions and velocities can conserve excess mass-energy as linear velocity. However due to the intrinsic oscillations of processor cycles allocated to computing velocity in space versus velocity in time those observable positions and velocities are inherently fuzzy with respect to each other and the reference background.

The virtual template by which dimensionality can be constructed can be visualized with one dimension suppressed as a plane x, y spatial surface rising at every recomputation in universal P-time. But because of the space versus time processor oscillations the surface of this reference template is oscillating between space and time differently for each computational process within the surface.

Thus when the exactly conserved values of energy are mapped to positions and velocities against the oscillating background the resulting values appear intrinsically uncertain with space and time conflated at the quantum scale. These oscillations are the source of the zero-point energy fluctuations of the quantum vacuum and the derived Uncertainty Principle.

How the oscillations produce the dimensionality of individual particles as they are computed can be described by wavefunctions but particle wavefunctions don't capture the essential coherence of all particles within the dimensional fragment. This is why the entanglement of particles created together and being computed as aspects of a single coherent process is a mystery in current quantum theory.

The oscillations are the reason that the exact trajectories and time evolution of individual entangled particles are describable as wavefunctions. However each *set* of entangled particles is computed as a separate coherent process by an individual application of the elemental program and processor. As a result the oscillations of each entangled *process* are unique and must be visualized as a separately oscillating plane. These separate oscillation planes are effectively fragments of dimensionality, each a separate variant of the virtual dimensional template with respect to which observable dimensional values are computed.

Then as particles from separate planes interact and decohere in new events they begin to be computed by new applications of the processor and form new dimensional fragments. At this point the previous dimensional fragments from which the interacting particles decohered are reduced to their remaining particles and continue their independent evolutions. In this manner entanglements are in a process of continual reduction and recreation and this overall process produces the observable dimensional values of the entanglement network we interpret as a physical spacetime.

In this model all the quantum equations still work as they did before, it's only the interpretation that is new. A wavefunction no longer represents a particle smeared out in a single fixed space, but the uncertainty of how a particle's dimensionality relates to the oscillating background of dimensionality itself as it's computed. Wavefunctions now become a description of the probabilities of how particles from separate dimensional fragments can join and decohere in an event common to both to produce a new dimensional fragment.

Thus the positions and velocities (momenta) of particles that are part of a single coherent process and single dimensional fragment are computationally exact with respect to one another, but are not exact with respect to particles of other processes being computed with different processor oscillations. However if two particles from different processes are to interact their positions and velocities must become exact with respect to each other for their energies and momenta to be exactly conserved.

This requires the positions and velocities of interacting particles to become momentarily exact with respect to each other. This results in a decoherence in which the dimensional fragments of the two particles transiently align and the decohered particles leave their previous dimensional fragments and begin to be computed as part of a new dimensional fragment by a new processor application with its own dimensional oscillations.

A decoherence is effectively a measurement of the position and momentum of each particle by the other and the results can be amplified to a lab instrument. This is the basis of all particle measurements, which are all decoherences of individual particle interactions amplified by measuring instruments or sensory systems.

So positions and velocities are exactly fixed among entangled particles exiting an event as a dimensional fragment, but not fixed with respect to particles exiting separate events in their own dimensional fragments. Thus if the positions and velocities of one particle in a two particle dimensional fragment is measured (within the limits of uncertainty) one would know the exact position and velocity of the other one at that same moment. Again within the limits of uncertainty, which is the underlying oscillation effect.

When decoherence occurs in an interaction between particles of two separate coherent systems it instantaneously aligns the two systems and all member particles. A decoherence of one entangled particle automatically decoheres the other. But as soon as the interaction event occurs it begins to be computed as a new independent process and the interacting particles lose their previous coherences with their parent dimensional fragments, which continue on in a reduced form on their own.

THE NATURE OF PROBABILITY FUNCTIONS

Ultimately a wavefunction is simply a probability function rather than anything 'real'. Contrary to the standard interpretation of quantum theory there is no physical wavefunction that particles are. Particles are exact *numeric* data trajectories whose *observable* dimensionality over the exact trajectory varies probabilistically due to the oscillations of the dimensional background against which observable dimensionality is computed.

Thus there were never any other actual dimensional values than those actually chosen even though they appear randomly and weren't themselves actual until they occurred. The other possibilities in the wavefunction have the same status as the possibilities of a thrown dice landing on a particular side. As soon as the dice lands on one particular side the other possibilities 'collapse' and vanish erased by the reality of the side landed. So the probability distributions described by wavefunctions were never anything real, but only an *observer description* of how the particle's dimensionality *might* become real. Nothing physical or real has happened to the other probabilities; it's just the description that has changed and all such descriptions exist only in observers' brains rather than in the universe itself.

If a man takes the train in the morning the probability that he didn't take the train vanishes to zero but there was never anything real or actual associated with that possibility. This is the essence of wavefunction 'collapse'. The wavefunction was never something physical or even numerically real that existed, it was simply some observer's probabilistic description of what *might* exist, and as actual events occur the descriptions of what might be continually change with no actual consequences at all. They were never part of reality in the first place but only some observer's description of a possible future reality that never had any actual reality.

Millions upon millions of probability descriptions of practically anything can be imagined with greater or lesser accuracy. But making one up doesn't add anything to the real world, which continues on its merry way. Same with wavefunctions, they are merely descriptions of how things might become observable; they are not real in any other sense at all so there is nothing actually collapsing but the description of an unfulfilled possibility. Wavefunctions are useful in making predictions, but only in the same sense as anything else. It's not the prediction that is real but the actual behavior of what was being predicted. The whole notion that wavefunctions are somehow actual particles is entirely misguided and a wholly incorrect interpretation of quantum reality.

This is also true about every aspect of the past including the complete fine-tuning. No matter how we can imagine any part or parts of it could have been different these are all completely imaginary *descriptions*, which never had any actual reality whatsoever. And every one of those possibilities has now collapsed into non-existence by the actual events of the past. Thus there never was any actual possibility of the past being even one iota different than it actually was and this includes the complete fine-tuning itself. There was never any actual possibility of the complete fine-tuning being different than it was and continues to be. All other possibilities are completely and absolutely falsified by the actual data state of the observable universe in the present.

One might claim the difference between classical and quantum predictions is that we are dealing with a complex probability amplitude in the form of a wave that can interfere with itself (as in the double slit experiment) and other predictions. But this is not true. Classical level predictions are frequently interdependent on each other. And when any prediction is itself complex with multiple interconnected aspects, changes in one aspect can often affect the others. So wavefunctions are not nearly as mysterious as they are made out to be.

So wavefunctions are not real, they are not part of computational reality itself, but in our interpretation they do reveal something important about how processor cycles oscillate. They oscillate in a manner whose probabilities are accurately describable by wavefunctions and the Schrödinger equation.

THE SOURCE OF QUANTUM RANDOMNESS

The elemental program computes everything exactly to the extent it has algorithms that cover them. But whenever the laws of nature have no exact decision making rules decisions are made randomly among possible choices based on their probability distributions. This is a basic principle of reality and the source of all the randomness in the universe and the reason reality is not deterministic.

Random choices happen when events could validly emit alternate particle sets, or when particles can be emitted with alternate trajectories while still conserving energy and momentum. In such cases dimensional values are randomly chosen from available choices according to their probabilities, which generally have to do with minimizing energy levels though in the case of spin entanglement the probabilities of any equal and opposite orientation are the same because there are no energy differences.

And importantly it's also true of the space versus time oscillations of the processor that computes the observable universe. Or in the usual quantum interpretation the measurement of observables from wavefunctions where observable values are chosen randomly from their probability distributions in a process of decoherence.

It's specifically true of the spin orientation paradox where the decision is delayed. In the case of spin orientation the randomness occurs not as the entangled spins are created but when they are measured, as they must decohere to exact values to be conserved by events.

For example, the spin orientation of the first particle measured can have any possible random direction within 3-dimensional space.

Whatever that random choice turns out to be the spin orientation of the other particle will automatically also decohere to be equal and opposite since that opposite spin relationship was already part of the two particle dimensional relationship, which is now aligned to the orientation of the laboratory dimensionality.

In general quantum events are random because there are no mathematical rules for how to exactly align the separate independent dimensional fragments produced by separate processes, so nature must choose alignments randomly within allowable constraints. This is effected by the randomness of the processor cycle oscillations.

Most randomness at the quantum level seems to occur with respect to dimensionality as it arises computationally in the merging of separate dimensional fragments as they are aligned and created by decoherence events. Thus prior to the emergence of dimensional interactions the computations of reality can be quite simple and exact as they contain none of the complexity necessary to describe events in relationship to intrinsically fuzzy dimensionality.

Ultimately the random oscillations of processor cycles as dimensionality is computed appears to be the source of nearly all the randomness of the universe, since all actual randomness occurs only at the quantum level. The apparently random processes of the classical level are either quantum level randomness amplified by supportive computational structures or simply processes that are too complex to be exactly computed. Most classical level non-predictability such as weather forecasts is a combination of the two.

This mechanism of random choice due to a lack of exact decision-making rules at the quantum level is the ultimate source of all the randomness of the universe. Without this single mechanism the universe would be completely deterministic, the future would be fixed, and there would be no free will. Thus all the randomness of reality which gives meaning and frees the universe from complete determinism is an emergent manifestation of the manner in which dimensionality is computed by the processor cycles that compute the universe. It's the quantum scale oscillations of the processor that frees us and the entire universe from complete determinism.

RESOLVING QUANTUM PARADOX

The common characteristic of most quantum 'paradoxes' is that they are paradoxical only with respect to a mistakenly assumed fixed pre-existing spacetime background, but in Universal Reality this fixed pre-existing spacetime doesn't exist so the paradoxical nature of quantum events vanishes.

Universal Reality treats spacetime as a computational structure that emerges at the level of particle events as frames of dimensional relationships. Because these dimensional relationships are directly computed by the conservation laws we know they exist and we know what they are. But there is no reasonable source in traditional science for the single universal spacetime that it assumes. How does it get there? Where does it come from? Traditional science has no answer.

Combine this with the fact that the empty spacetime of traditional science is not even observable, that only individual dimensional relationships are actually observable. A general principle of science is that unless something is observable its reality is questionable at best.

Thus it's eminently reasonable to treat spacetime as independently oscillating frames of dimensional relationships computed by quantum events. Not only is this approach much more reasonable but it resolves the apparent paradoxes of quantum theory in a completely natural non-paradoxical manner as the alignments of separate dimensional frames as they are joined by decoherence events. An excellent example is how it resolves the spin entanglement paradox as described above.

The model of entangled particles as dimensional fragments computed as independent processes neatly resolves the seemingly paradoxical nature of quantum phenomenon. All quantum paradox is now seen as an artifact of science's mistake of assuming an exact fixed preexisting spacetime container within which events occur. This is due to a lack of understanding on the part of scientists of how our minds create a physical spacetime reality that is entirely different than actual reality.

We have already explained the source of quantum randomness, the wave-particle duality of nature, the Uncertainty Principle, and the spin entanglement 'paradox' in a non-paradoxical manner but it's worthwhile covering a few more aspects of quantum phenomena to really clinch the case.

For example in the case of the double slit paradox that is often taken to demonstrate the wave nature of material particles it's not particle waves that are interacting with the slits in a fixed space but oscillations of the dimensional background with respect to which the positions and velocities of the particles become observable. It just so happens that the wavefunction model properly describes the form of these oscillations, though the interpretation is reversed.

In this model the slit and screen apparatus itself consists of molecular structures composed of bound particle entanglements (see the eponymous section below) so the dimensionality of the apparatus is more classically constrained.

In terms of the oscillating background model the dimensionality of the apparatus is oscillating relative to the exact trajectory of the particle, at least that's the way it appears to the particle. So the observational effect is the same. From the perspective of an observer conducting the experiment the particle appears wave-like, but from the perspective of the particle the observer and apparatus appear to be oscillating in waves of inverse form.

Then as the particle hits the screen it decoheres with the dimensionality of the apparatus and observer (within their own limits of uncertainty and bound entanglement) and becomes part of that system and begins to be computed as part of it.

This is consistent with our supposition that the supposed wave nature of particles is not an observable phenomenon and thus suspect. Only the particle impacts on the screen are actually observable. The wave nature of particles is never directly observed but only *inferred* from the *pattern* of impacts. But our interpretation in which processor cycle oscillations produce observable dimensionality explains these patterns perfectly well without wavefunctions.

When a particle is measured to determine which of the two slits it passes through the interference pattern on the back screen disappears. In our model the measurement decoheres the particle and it becomes part of the double slit system at that point and its coherent trajectory begins again from that location so the probability of its dimensionality oscillating through both slits has vanished to zero. All other apparent quantum paradoxes are explained in a similar manner by our model.

WAVEFUNCTIONS

Universal Reality can be considered a new interpretation of quantum theory that emerges naturally from the much wider perspective of its comprehensive Theory of Everything. But Universal Reality doesn't try to replace the equations that are the actual essence of quantum science. These equations are wonderfully accurate and comprehensive and are the cumulative work of a century of great genius.

Universal Reality accepts these equations because they work, but it completely revolutionizes our understanding of what they are telling us, and the overall context in which they have meaning. It completely changes their *interpretation*.

For example wavefunctions are the primary mathematical description of quantum entities and they describe their behavior with great accuracy, yet even physicists admit they don't understand them. Universal Reality provides a context in which wavefunctions naturally acquire a straightforward and non-paradoxical meaning.

In Universal Reality wavefunctions still work, but they aren't descriptions of particles with respect to a fixed pre-existing spacetime background, but descriptions of how individual particles in dimensional frames are formed, joined and aligned by quantum events. Thus wavefunctions don't describe particles that are smeared out in space and somehow everywhere at once, but how the separate dimensional frames associated with particles oscillate with respect to each other. The mathematics is the same but the interpretation is much more natural and reasonable and completely non-paradoxical as well.

In Universal Reality particles are exact computational data entities that always act as particles. This is true because only the particle nature of particles is ever directly measured or observed. Their apparent wave nature is always deduced from patterns of individual measurements *of particles acting as particles*. Because particle waves are never directly observed their existence must be considered questionable. Wavefunctions don't describe the wave nature of individual particles, but the randomness in the way their dimensional relationships can decohere with respect to the frame of an observer or other particle.

This resolves the fundamental quantum paradox of the apparent dual nature of particles as particles and wavefunctions. Particles are always particles with exact trajectories and time evolutions as their

Schrödinger equations imply. Their apparent wave like nature has nothing to do with the particle itself but with how its dimensional observables are intrinsically uncertain due to the continual oscillations of the reference background with respect to which its dimensionality is computed. It's dimensionality itself that is fuzzy as it's computed rather than particles fuzzy with respect to an exact dimensional spacetime.

QUANTUM TUNNELING

Quantum tunneling is a seemingly paradoxical phenomenon in which particles sometimes appear on the other side of an impenetrable barrier (Wikipedia, Quantum tunneling). The standard interpretation of tunneling in quantum theory is that the wavefunction of the particle is smeared out in space and that part of it extends to the other side of the barrier. And since the wavefunction is the probability distribution of where the particle might appear it can sometimes appear beyond the barrier.

Universal Reality reinterprets quantum tunneling in terms of its computational model. Here again it's not the particle that is smeared out in an exact fixed space, but dimensionality itself that is fundamentally oscillating as particle positions and velocities are computed and pinned against it to produce observable values. Because they are being computed by separate applications of the processor their dimensionalities inherently oscillate with respect to each other and the resulting fuzzy dimensionality of the particle with respect to the barrier is describable by a wavefunction. This is fine so long as the wavefunction is understood as an abstract observer description rather than the actual particle.

So quantum tunneling is due to probability distributions not of particles but of the underlying dimensionality against which they are pinned. There is no physical spacetime in which objects have positions. The universe is all numeric, so there is no intrinsic problem of things appearing on the other side of things. It's just a matter of consistently computing observable positions and velocities.

The particle and the barrier are being computed as different processes and the processor cycles computing them have different oscillations with respect to each other. Thus if their oscillations overlap sufficiently the particle has some probability of appearing on the opposite side of the barrier.

The important point here is that each coherent process is computed by separate applications of the processor, each with its own oscillatory pattern. There is no single physical spacetime background each of whose points is characterized by a particular oscillation or lack of oscillations. The oscillation patterns are particular to separate processor applications within non-dimensional computational space. Points of the reference background can be visualized as oscillating differently in each process. The oscillations are different in each process because each is calculated by a separate application of the processor, and the computational results of the oscillatory pattern also depends partly on the individual details of what is being computed.

More massive mass-energy structures are computed with smaller dimensional oscillations. Quantum theory explains they have shorter de Broglie wavelengths, thus their oscillations are less than those of free particles (Wikipedia, Matter wave). As the mass and size of a structure increases it behaves more and more classically and impenetrable barriers are not penetrated.

HALF-LIVES

Half-lives are another good example of the seemingly paradoxical nature of quantum reality. There are two types of half-lives of interest (Wikipedia, Half-life).

Every radioactive isotope has a half-life that describes how long it takes for half of its nuclei to split and decay. The decay of any particular nucleus appears to occur completely randomly, but in aggregates of many atoms approximately half will decay in every half-life interval specific to that isotope.

In our model decay is a matter of certain dimensional adjacencies occurring. The internal elements of the nucleus are all oscillating randomly with respect to each other by a minute amount and when the overlap is sufficient to overcome the prevailing balance of forces holding the elements together a decay results.

If a fissionable nucleus has internal components whose dimensionality jostles probabilistically with respect to each other they can come close enough to react on a random basis. They must be closer than some threshold distance from one another to interact for the nucleus to split. Thus it's the dimensional oscillations that bring them close enough for a decay to occur but with a predictable frequency in aggregate based on the particular details of the oscillations of the processor application computing them.

In nuclei this is due to the mutual jostling of the quarks and gluons in the nucleus, which are in a constant exchange that holds the nucleus and its protons and neutrons together. When this balance is randomly upset the nucleus splits. Everything jostles in just the right manner to hold the nucleus together until the jostling randomly exceeds one of its probability thresholds and the nucleus splits. The average statistical form of the jostling determines the aggregate half-life but its random nature determines the fission of any particular nucleus.

The other type of half-life decay is that of individual particles. For example free neutrons decay into protons, electrons and neutrinos with a half-life of around 10.2 minutes (Wikipedia, Neutrons). Many other particles including various mesons have much shorter half-lives and decay almost instantly. Basically only particles that are composed of multiple other particles in the sense their particle components can be validly redistributed into other particles are subject to decay and have half-lives. However free protons, which are composed of quarks like neutrons are not known to decay. Particle decay is essentially an event involving a single initial particle and is computed similarly to multi-particle nuclei.

Particle decay, like most processes in nature, is generally a mechanism to reach a lower energy state as some of the particle's internal energy is converted into the kinetic energy of its resulting particles.

Thus the probabilistic half-lives of composite particles and nuclei arise from the particular oscillations of the internal details of particles subject to decay relative to each other. Decay will result when particular overlapping oscillations occur and those occur probabilistically based on the oscillations computing the internal structure just as with radioactive nuclei.

Thus the mechanism responsible for half-lives is the same underlying mechanism of processor oscillations responsible for all quantum phenomenon and their seemingly paradoxical nature. Again the process is similar to that of quantum theory but the interpretation is reversed.

BOUND PARTICLES

Up till now we have been concerned with free particles but of course it is particles bound in atoms and molecules that make up most of the structure of the universe.

Bound particles are essentially bound entanglements of continuous interactions, and form a major part of the entanglement network. It appears that all the complexities of atomic and molecular structures emerge naturally from the simple rules that govern the interactions of free particles. Thus atomic and molecular matter is actually an emergent phenomenon that arises naturally from inter-particle interactions according to the rules of the complete fine-tuning.

This means that all of the very complex equations that describe atomic and molecular matter are not part of the programs that actually compute the universe. These equations are *descriptions* of the aggregate behavior of the elemental program that actually computes particle events. All the emergent structure of the universe is simply the manifestation of individual particle interactions at the aggregate level.

There appear to be no higher level laws involved in actually computing the universe, though the programs of purposeful beings appear to act as such in the same manner that complex computer programs are directed towards computing various tasks but always only in terms of actual machine level operations. This is a complex subject that is explored further in the chapter on Emergence in Universal Reality (Owen, 2016).

The atoms and molecules that make up all the mass-energy structures of the universe are composed of bound particles, specifically the electrons that occupy orbitals around the protons and neutrons of nuclei, which in turn are composed of quarks bound by the strong force by gluons. The rules that govern particle binding are consequences of a few simple rules that govern how different types of free particles and their particle components interact.

In atoms, electrons are attracted to the opposite charge of protons in the nucleus and drawn towards them, however in most cases they don't have enough energy to react with a proton to produce a neutron. The mass of a neutron is significantly greater than that of a proton and electron combined. So for a reaction to occur there must be enough additional energy to be converted into the additional mass necessary to form a neutron. This can only occur with very high velocities or intense gravitational fields that provide enough intrinsic velocity energy to be converted into the necessary additional mass.

This does occur in extreme cases such as neutron stars where atoms are crushed by intense gravity and electrons do combine with protons to form neutrons and all the atoms of the star collapse to the size and density of their nuclei.

Thankfully in most cases electrons don't carry enough energy to react with protons to form neutrons, and electrons in atoms are unable to react with their protons. Thus it's only the slight mass disparity between neutrons compared to protons plus electrons that prevents all atoms from collapsing into neutrons and all the ordinary matter in the universe from disappearing! An important example of how the universe is very finely tuned to maintain its structure.

Instead electrons are trapped in atoms by nuclear protons and continually bounce back and forth around them because they are unable to react with them or escape from them. The electrons become bound by the electrostatic attraction of protons and oscillate around the nucleus forming standing waves of the probability distribution of where the electron might be at any moment with respect to the nucleus.

The forms bound electron waves take around nuclei are called atomic orbitals. The basic principles underlying the forms of atomic orbital are fairly simple through the actual resulting forms in multi-electron atoms become quite complex due to electron-electron interactions and the imperfect spherical attraction of multi-proton nuclei (Wikipedia, Atomic orbital).

Because they are constrained by the attraction of nuclear protons bound electrons settle into harmonic standing waves centered on the nucleus. Harmonic waves are standing waves with an integer number of nodes that maintain their forms over time. They are analogous to a violin string fixed at both ends, which can only vibrate in a standing wave of one, two, three, or more integer numbers of nodes when plucked.

Because electrons form standing waves their energies become fixed. Each different standing wave has a specific energy and for an electron to jump between from one wave form to another it must absorb or emit a specific amount of energy in the form of a photon. The specific frequencies of the photons emitted or absorbed accounts for the distinctive spectral colors and lines of the various elements, and the fact that atomic orbitals have discrete energies is the origin of quantum theory.

The quantized energy levels result from the relation between a particle's energy and its wavelength. For a confined particle such as an electron in an atom, the wavefunction takes the forms of standing waves. Only stationary states with energies corresponding to integral numbers of wavelengths can exist; for other states the waves interfere destructively, resulting in zero probability density.

A more accurate analogy is that of a circular drum head whose circumference is fixed to the drum rim. Depending on how it's struck it vibrates as a standing wave with one or more nodes and the waves produced in the drumhead are nearly identical in form to those of electron orbitals. There are some excellent animations illustrating this on Wikipedia (Wikipedia, Atomic orbital).

So the secret to understanding electron orbitals is they are all the possible modes of dimensional oscillations of standing waves with increasing numbers of nodes in 3-space around a center that constrains them. This is the simple key to understanding electron orbitals and underlies the periodic table of elements.

Based on this principle atomic orbitals form successive shells of increasing radius around the nucleus. Orbitals in an atom are uniquely described by 4 quantum numbers so that no two electrons in an atom can have the same 4 quantum numbers. The first quantum number n denotes the shell and is simply the number of nodes electron waves in that shell have; 1, 2, 3, etc.

The second quantum number l is the azimuthal quantum number and ranges across all integer numbers such that $0 \leq l \leq n-1$. Thus for each quantum number n there is a set of $l+1$ quantum numbers corresponding to the number of possible harmonic wave forms with n nodes. The azimuthal quantum number basically describes the orientations of possible standing wave forms relative to the 3 spatial axes. There are 3 possible identical harmonic waveforms, one along each axis.

The 3rd quantum number m_l, the magnetic quantum number, describes the magnetic moment of an electron in an arbitrary orientation and is also an integer that varies within the subshell l_0 such that $-l_0 \leq m_l \leq l_0$. So for example for subshell $l=2$, m_l would take on the values -2, -1, 0, 1, 2 corresponding to the possible harmonic wave forms for that n shell and *l* subshell. All the *m*'s within all *l*'s for a given n correspond to all the possible harmonic wave forms with n nodes along all 3 coordinate axes, and together for all n's these define all possible orbitals for an atom.

The number of possible waveforms a spherically centered standing wave of n nodes can have in 3-space is n^2. However two electrons can assume the same orbital waveform if their spins are oppositely oriented so the maximum number of electrons in a shell becomes $2n^2$. Spin, s, is the last of the 4 quantum numbers and is always plus or minus ½ (spin up or spin down) since the electron's spin is ½.

Thus the possible orbitals electrons can occupy in an atom is simply the number of possible harmonic waveforms the electron wave can symmetrically assume centered on the nucleus.

The orbitals are the waveforms that actual electrons must assume when they are filled. The actual number of electrons in a neutral atom is equal to the number of protons so the charges are balanced, and the number of protons determines the element.

In turn atoms form molecules primarily by sharing outer orbital electrons. When atoms combine in molecules outer atomic orbitals become distorted into molecular orbitals and the specifics of how these form and their complex properties determine the laws of chemistry and thus the structure and interaction rules of all matter. Thus chemistry and all mass-energy structures are the emergent results of the bound interactions of electrons, protons and neutrons.

A somewhat analogous situation occurs in the nucleus where the mutual repulsion of protons is overcome by the strong force, and proton and neutron quark waves vibrate in complex standing wave forms around each other. However, because the masses of protons and neutrons are much greater than the electron mass their de Broglie waves are much smaller, and thus nuclei are much smaller than electron orbitals.

The orbital forms shown in most illustrations are those of ideal individual harmonic waves. They are those a single electron would assume around a single proton as it increased or decreased its energy and jumped from orbital to orbital. However the presence of multiple electrons occupying orbitals in a single atom distorts their orbital forms due to the mutual repulsion of electrons and the necessary presence of multiple protons so the positive charge of the nucleus is no longer exactly spherically symmetric.

Thus though the principles underlying electron orbitals are fairly simple the forms they actually take in atoms become quite complex with increasing atomic number due to the mutual repulsion among electrons, including the screening of positive nuclear charges from electrons in further out shells, the uneven attraction of multiple protons in the nucleus, and even relativistic effects coming into play in outer orbitals.

In more technical terms the electron wavefunction oscillates around the nucleus according to a time independent (unchanging) Schrödinger wave equation, and orbitals are its standing waves. The standing wave frequency is proportional to the orbital's kinetic energy. The real part of the Schrödinger equation gives the form of the orbital, and the imaginary part gives the probability distribution of finding an electron at a particular location within it.

Thus emergence begins at the atomic level more or less. Atomic and molecular structures emerge automatically from the way that electrons, protons and neutrons interact, from the computational rules that actually compute their interactions.

Electron orbitals, and thus all chemistry and the structure of all matter emerges from the possible harmonic forms bound electron waves can take around protons, and electron waves themselves are the result of the space time oscillations in the processor cycles that compute reality.

Thus a small set of fairly simple particle interaction rules produces the atoms and molecules that compose all material structures:

1. The mutual attraction and repulsion of electrons and protons.
2. The inability of electrons to react with protons at normal energies.
3. The resulting possible orbital forms of standing harmonic electron waves centered on the nucleus.
4. The analogous strong force binding of proton and neutron quarks in nuclei.

Thus the structure of the observable universe from atoms on up is emergent rather than independently computed. All the actual computations of reality occur at the particle and particle component level, and larger scale structures automatically emerge from bound particles that manifest these elemental computations. This is how all the incredible complexity of the observable universe emerges from the finely tuned interplay of the simple elemental rules that actually compute them. This greatly simplifies the computational structure of reality. It's the incredibly amazing complete fine-tuning that is responsible for the wonderfully meaningful complexity of emergent structures that are simply its aggregate manifestations.

Because they exist as bound entanglements the dimensional relationships of electrons and protons are exact in terms of energy conservation, which is of course the basis of quantum theory. However when these conserved energy relationships are viewed in terms of positions and velocities of electrons relative to the nucleus they are subject to the processor cycle oscillations as viewed from an observer frame, which is why they appear as harmonic standing wavefunctions within which the actual position and velocity of the electron appears probabilistic.

MASS-ENERGY STRUCTURES

The emergent observable universe consists of the data of particles and their fields in a dynamic balance of the four fundamental forces. All the atomic and molecular structures of particles are balances of different types of forces. Just as all forms of mass-energy are different types of relative motion, emergent particle structures form when different forms of relative motion reach balances.

This is due to the attraction and repulsion of the various forces on the various charges. The charges of the forces manifest fields that attract and repel other charges. Due to the principle of differentiation, the universe consists of myriads of discrete particles with different types and signs of charges. Thus the fields of the various forces balance to form compound particle structures, and these balances define the atomic and molecular structures of the observable universe.

Atoms are balances of the repulsion and attraction of electric charges, and atomic nuclei the balance of repulsion and attraction of the strong force operating via gluons on the color charges of constituent quarks. And the gross structure of the cosmos is largely a balance of the gravitational attraction of mass charges with the linear relative motion of kinetic energy in combination with the repulsive gravitation of the Hubble expansion.

All the 4 forces can be viewed either as forms of relative motion or as exchanges of particles due to the equivalence of space and mass-energy as forms of spatial velocity. Basically the emergent structure of the observable universe is a balance of the relative velocities of the 4 forces with the relative velocity of linear motion. Everything is given its emergent structure by the interplay of various types of relative motion computed at the particle level.

Because spacetime and particle structures are computed together as an integrated structure, and because atomic and molecular structures are so pervasive, they are the main contributor to the overall structure of the observable universe. They are the primary constituents of the entanglement network at least in the vicinity of stars and planets.

To summarize Universal Reality's model of the universe unifies quantum theory and general relativity in a single computational model in which relativistic spacetime emerges from quantum events. Particles are still well described by the equations of quantum theory as are large-scale phenomenon by the equations of general relativity. Universal Reality just interprets them differently, and this new interpretation leads directly to their unification and numerous new insights into the ultimate nature of reality.

EPILOGUE - TESTING THE THEORY

Every new theory must be subject to experimental tests to either confirm or falsify it. For example Einstein's seemingly outlandish relativity theory would never have been accepted had it not made testable predictions of the bending of starlight by the sun's gravitational field (Eddington, 1928). In this case the tests were straightforward, Einstein was confirmed, and relativity quickly became an accepted theory.

Other theories have not been so lucky. For example the theories of evolution and of plate tectonics languished for years without simple conclusive tests until gradually the evidence became overwhelming. This is the usual case. Science progresses slowly and carefully but in the end it always progresses.

The theory of Universal Reality faces the same challenges. Since it's mostly a completely new reinterpretation of accepted science in the framework of a much broader Theory of Everything it's difficult to isolate clear tests that could either confirm or falsify it. However a few possibilities do come to mind.

Hopefully there will be at least some who test the theory against both experimental evidence and by consistency with the general body of scientific knowledge. However one must always be careful to test against actual mathematical theories rather than current *interpretations* of scientific theory that are not part of the science proper. These tests may involve trying to find reasons why Universal Reality can't possibly work, and those will be useful in bringing to light points that need development, but I suggest the more fruitful approach with any new theory is to try to find ways to make it work.

The theory of Universal Reality appears to have much promise in that it explains so much so well from a single universal approach. The general approach has been to accept the experimentally confirmed equations of science but develop a completely novel unifying interpretation that incorporates all aspects of reality in a single Theory of Everything. The author believes Universal Reality is the best most comprehensive Theory of Everything on the market and urges others to put it to the test and report their findings to Edgar@EdgarLOwen.com.

Since much of Universal Reality is a new interpretation of science and other aspects of reality it may not be subject to experimental tests.

However it can be tested with respect to its logical consistency with accepted science. Overall consistency across all aspects of reality is the true and ultimately only test of validity.

There are other useful tests as well such as elegance, simplicity and beauty. Universal Reality is founded on a set of simple principles of universal scope, and scores high on these criteria. And it proposes a computational model that is quite parsimonious compared with many of the currently fashionable interpretations of reality.

Not only does it explain the universe of science quite well but it does do in a manner that intuitively integrates existence, consciousness, and the present moment, the fundamental experiential constituents of reality about which current science has had nothing meaningful to say.

There are a number of specific testable proposals of Universal Reality that come to mind and there are no doubt others. Proving the existence of a universal present moment common to all observers and processes across the universe is a crucial one. This should be easy to confirm by simply proving there is a unique one-to-one correspondence between the proper times of all clocks upon which all observers agree.

Also of fundamental importance is a mathematical confirmation of the theory's core notion that a spacetime that is computationally created along with mass-energy structures by quantum events is key to unifying quantum theory and general relativity. To anyone interested in unifying relativity and quantum theory I strongly suggest this is the correct approach.

Confirming a slightly positive curvature to space would tend to confirm Universal Reality's prediction that the universe has a closed finite positively curved hyperspherical geometry.

Universal Reality's theory of Dark matter as a spacetime curvature produced by the predicted uneven Hubble expansion of space around the edges of galaxies and galaxy clusters should be confirmable by measurements of the strength and distribution of dark matter relative to the motion of the galactic masses that produced it.

Any detection of dimensional drift or other relativistic anomalies would tend to confirm Universal Reality's theory of a computationally based absolute spacetime background with respect to which rotation and world lines are relative. In fact confirmation of the necessity of an absolute dimensional background in relativity itself will also confirm Universal Reality.

There are likely testable consequences of Universal Reality's theory of dimensional fragments. All entangled particles should have exact instantaneous dimensional values with respect to each other at least within the limits of uncertainty that evolve in a predictable manner and may be within the range of measurement. In other words the wavefunctions of entangled particles should be coherently coupled and this coherence may be observable.

There are also many other proposals of Universal Reality that are potentially subject to experimental and perhaps theoretical falsification. The inability to falsify these would lend considerable credence to the theory. A few are of particular importance.

The METc Principle that all forms of mass and energy can be consistently modeled as different forms of spatial velocity.

And lastly the ability to program a convincing simulation of reality on the basis of the theory of Universal Reality that correctly models the major known aspects of reality would be the best test of all and a very strong confirmation of the theory. We have already taken some initial steps in this direction that seem quite promising so far.

NOTES

1. DERIVING RELATIVITY FROM THE STc Principle

For the mathematically inclined the Lorentz transformation that governs most of special relativity can be derived from the STc Principle as follows indicating the STc Principle underlies special relativity. To simplify the discussion we assume that any spatial velocity is parallel to the x-axis. Then the STc Principle can be expressed mathematically as

$$\mathbf{v}_x + \mathbf{v}_T = \mathbf{c} \quad (1.1)$$

Where $\mathbf{v}_x$ is the velocity through space along the x-axis, $\mathbf{v}_T$ is the velocity through time, and $\mathbf{c}$ is the velocity of light. Writing the quantities in bold indicates that they are vectors; that is they have both a magnitude and a direction. Note that velocity in time is actually the relative time rate of any clock to an observer's clock multiplied by the speed of light c to put it in the same units as spatial velocity so time can be treated as another dimension of a consistent 4-dimensional geometry.

Expressing the x velocity as a vector $\mathbf{v}_x$ is standard physics, however some physicists might recoil at expressing the velocity of time as a vector since it's normally considered to be a scalar, a quantity having a magnitude but no direction. In fact though, in a 4-dimensional universe, time clearly does have a direction along the time axis, and thus is most certainly a vector. The addition of two vectors always produces another vector, so that the result of Eq. (1.1) is a vector *velocity* in a particular direction and the magnitude of that velocity is always the scalar *speed* of light.

Eq. (1.1) can be depicted graphically. In the following figure the vertical axis is the velocity of time, and the horizontal axis is the velocity in space along the x-axis. When we plot Eq. (1.1), we get a circle of constant radius c = the speed of light. The graph shows an arbitrary example whose spacetime velocity vector extends from the origin to the circle along with its space and time velocity components projected on their respective axes. It also shows the cases in which the c spacetime velocity is directed either entirely along the vertical time or horizontal space axis.

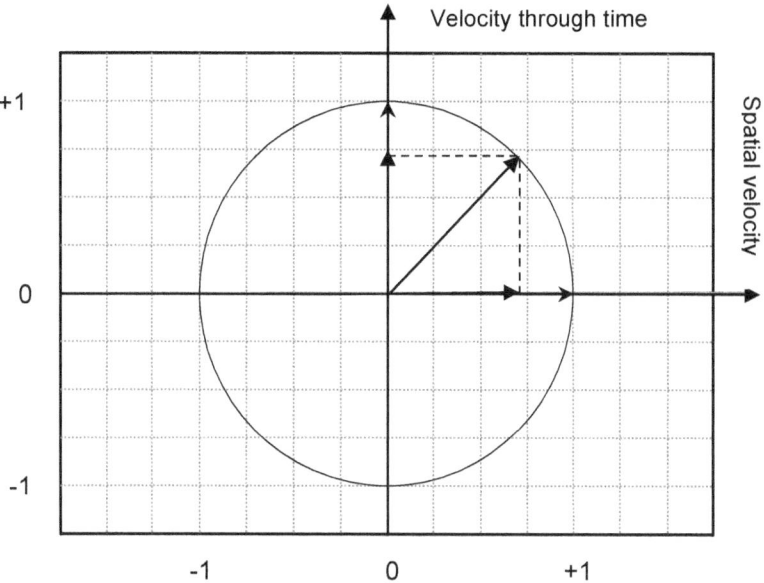

Fig. 1.1. The STc Principle. The vector sum of space and time velocities of every entity in the universe is always equal in magnitude to the speed of light c at every moment, as shown by the circle. Arrows show the three possible cases. When there is no relative velocity in space all an object's spacetime velocity is through time at the speed of light as shown by the vertical arrow touching the circle. In contrast, light itself always travels with relative velocity in space $v_x = c$ as represented by the horizontal arrow touching the circle. In the third case, when $0 < v_x < c$ the spacetime velocity is the vector sum of the projections of the space and time velocities along their respective axes as shown by the diagonal arrow. But in all cases the vector sum of space and time velocities is equal in magnitude to the speed of light as indicated by the circle. Though the c velocity is represented as a complete circle in the figure only velocities in the upper half circle where time is positive are actually possible.

The mathematical rules of vector addition are just expressions of simple geometry and the Pythagorean theorem so that we can alternately express Eq. (1.1) as

$$v_x^2 + v_T^2 = c^2 \qquad (1.2)$$

That is the sum of the squares of the space and time velocities is equal to the square of the speed of light. This is the more usual mathematical form equivalent to the vector form of (1.1).

From Newtonian physics we know that a velocity is the time rate of change of a distance so that the space velocity along the x axis could also have been written as dx/dt, the instantaneous rate of change of distance x with time t at any given moment.

Now what is meant by the velocity of time? The idea is that a moving clock runs at a slower rate than a stationary clock. So it makes sense to consider the rate at which the moving clock runs relative to an observer's stationary clock as the relative velocity of the time displayed by the moving clock. We express this velocity as dT/dt, the time rate of change of the moving clock as measured by the stationary clock.

Another consideration is that distance in space is measured in spatial units such as meters, whereas distance along the time axis is measured in time units such as seconds. To have a meaningful equation in which both types of distances and their velocities can be compared they must both be expressed in the same units. This can be done by expressing both in a unit such as light seconds, the distance that light travels in a second, $\approx 3 \times 10^8$ meters. This is easily accomplished by substituting cT for T to express the distance along the time axis measured by the moving clock. This gives the equivalent time distance (seconds) in units of spatial distance (light seconds). With these additions we can now rewrite Eq. (1.2) as

$$v_x^2 + \left(\frac{dcT}{dt}\right)^2 = c^2$$

and since for a constant c, dcT/dt = cdT/dt

$$v_x^2 + c^2\left(\frac{dT}{dt}\right)^2 = c^2 \tag{1.3}$$

Rearranging and dividing both sides by c^2

$$\left(\frac{dT}{dt}\right)^2 = \frac{c^2 - v_x^2}{c^2} = 1 - \frac{v_x^2}{c^2}$$

and taking the square root of both sides we have

$$\frac{dT}{dt} = \sqrt{1 - \frac{v_x^2}{c^2}} \tag{1.4}$$

which gives the velocity of time (the relative slowing of its clock rate) for a clock moving at velocity v_x relative to the observer's clock.

Now Eq. (1.4) is a fundamental equation of special relativity, an expression of the Lorentz transform, and from it all the standard effects of special relativity can be derived except for $E = mc^2$ which follows with the addition of the standard classical principles of conservation of energy and momentum. Thus we find that the theory of special relativity is in fact a consequence of the STc Principle.

2. THE BLOCK TIME DELUSION

Some physicists deny the concept of a present moment and instead believe in the block time theory in which all times exist at once and there is no preferred present moment (Wikipedia, Eternalism (philosophy of time)). This is a fundamental error based on a mistaken interpretation of relativity.

Relativity clearly reveals the universe as a 4-dimensional structure but that doesn't require the time dimension to have any actual extension. All we need to satisfy relativity is a 4-dimensional surface where the only moment in time that exists is the current universal present moment in which clock times are being computed.

The standard block universe model envisions the total past present and future history of the universe existing as a single fixed completely static 4-dimensional structure in some sort of vaguely defined eternal time or outside of time. Our apparent existence in this particular present moment we seem to experience is an illusion because we are also continually having the same illusory experience in every point of time of our block time world line corresponding to every static instant of our life.

In a block time universe the entire history of the universe already exists. The past still exists and the future is already fixed and exists as well. Thus it's completely fixed and deterministic and can never be any different than it is or changed it the least because change doesn't exist.

There are so many things wrong with this theory it's hard to know where to begin but we will mention the main ones.

1. It's contradicted by relativity itself since the STc Principle clearly demonstrates that everything in the universe has velocity in time, which necessitates it being at one privileged location in time at

any moment and that is where the present moment is. Thus contrary to the beliefs of some physicists relativity doesn't even work without a present moment.
2. In a block time universe the entire history of the universe already exists. The past still exists and the future is already fixed. Thus it's completely deterministic and fixed and can never be any different than it is. There is no randomness, chance or freedom whatsoever. Life is essentially meaningless.
3. In a block universe we are simply a sequence of static slices of our time line like stills from a movie and each of those static slices is having the illusion it's in a present moment. But if nothing is ever actually changing there is no way whatsoever that what is effectively a still photo of ourselves could be having any experiences at all. In short consciousness itself is impossible in a block universe because the experience of consciousness is an experience of something happening.
4. Block universe advocates tell us that the us and them that seem to be conversing right now are just static time slices of us both but there is no explanation whatsoever for why it's these particular static time slices that have been selected to do the talking. Somehow all our static time slices are always doing their thing but there is no attempt to explain why 'we' are the slice that we actually experience, nor is there any attempt to explain the obvious temporal progression from slice to slice our experience follows.
5. There is also the fundamental problem of how a block universe could come into being. It obviously has an enormously complex and consistent causal structure. How could such a structure be non-sequentially created *ab initio*? It seems to suggest the necessity of an omniscient creator god that stands outside of time, which just begs the question of how he was created. How would whatever process created it know how to create each state so that it would be completely causally consistent with the preceding one if the previous state didn't actually cause the successor state? But if previous states were the source of successor states then all states had to have been created sequentially and that is precisely the standard time evolving universe that the block universe denies!
6. A block time universe violates practically every scientific law known. For example it violates the conservation of mass-energy since the mass-energy of all time slices exists at once so the total mass-energy of a block universe is the sum across all time slices and the number of time slices could vary depending on the universe. But if we claim the conservation of mass-energy applies

only to single time slices then why and why is it consistent from slice to slice?
7. It's also completely impossible to speak meaningfully about science and the universe from the perspective of belief in a block universe. No matter how hard they try even the firmest believers must immediately revert to describing any aspect of reality or daily life in terms of clock time flowing through a present moment. Thus reality just doesn't make sense from the perspective of block time and this lesson should be taken to heart.
8. Finally advocates of the block universe theory tend to be true believers and take it as a tenant of quasi-religious belief. Any logical or scientific arguments raised against it tend to be summary dismissed as attacks on tenants of faith with no cogent reason given to reject them other than it just isn't so and doesn't really work that way.

In summary the block time universe theory is total nonsense that only a physicist could have invented and must be summarily rejected as the very obvious existence of an actual present moment and the flow of clock time completely falsify it.

BIBLIOGRAPHY

Eddington, Arthur Stanley. *The Nature of the Physical World.* 1928.
Goffinet, François, *A bottom-up approach to fermion masses.* Université catholique de Louvain, 2008.
Greene, Brian. *The Elegant Universe*. Norton, 1999.
Greene, Brian. *The Fabric of The Cosmos.* Vintage Books, 2005.
Halpern, Paul & Wesson, Paul. *Brave New Universe*. Joseph Henry, 2006.
Hawking, Stephen W. *A Brief History of Time.* Bantam Books. 1998.
Hofstadter, Douglas R. *Gödel, Escher, Bach*. Vintage, 1980.
Misner, Charles W.; Thorne, Kip S.; Wheeler, Archibald. *Gravitation.* Freeman, 1973.
Owen, Edgar L. *Spacetime and Consciousness*. EdgarLOwen.info. 2007.
Owen, Edgar L. *Mind and Reality*. EdgarLOwen.info. 2009.
Owen, Edgar L. *Reality.* Amazon.com. 2013.
Penrose, Roger. *The Road to Reality*. Knopf, 2005.
Price, Huw. *Time's Arrow and Archimedes' Point*. Oxford, 1996.
Schroeder, Daniel V. Purcell Simplified, http://physics.weber.edu/schroeder/mrr/MRRtalk.html, 1999.
Susskind, Leonard. *The Cosmic Landscape.* Little Brown. 2006.
Wigner, Eugene. *The Unreasonable Effectiveness of Mathematics in the Natural Sciences.* John Wiley, 1960.
Wikipedia contributors. *Wikipedia, the Free Encyclopedia.* http://wikipedia.org

Edgar L. Owen was born April 1st, 1941 and quickly realized that reality is not as it appears to be. A child prodigy, he entered the University of Tulsa aged 15 and received a B.S. with honors in science and mathematics with a minor in philosophy at 18 before completing several more years of graduate study in physics and philosophy.

In the early 60's he moved to the Haight-Ashbury in San Francisco where he hung out with notables from the Beat Generation, and conducted an intense personal study of the nature of mind and consciousness. From there he traveled to Japan where he lived for three years studying Zen and Buddhist philosophy while subsisting as a ronin English teacher.

Upon returning to the US he began a career in computer science writing numerous programs in artificial intelligence, simulations, graphics, and cellular automata while designing and managing advanced computer systems for the New York Federal Reserve Bank and AT&T. He then left the corporate world to start his own software business marketing his own CAD programs, which he ran for a number of years. Currently he owns a premier Internet gallery of fine Ancient Art and Classical Numismatics at EdgarLOwen.com.

Deeply immersed in nature since childhood, and always considering it the ultimate source of his inspiration and knowledge of reality, he has served as Chairman of his local Environmental Commission and organized several campaigns to protect the local environment and its wildlife.

Over the last several years he has worked to combine and organize the results of a lifetime of study of the various aspects of reality into a single coherent Theory of Everything. He now spends most of his time exploring the wonderful awesome mystery of reality and how it can be experienced more fully and deeply and enjoying his existence within it.

Edgar currently lives in Northern NJ in a big old house on top of a hill where he communes with nature and enjoys the company of his wild visitors including the occasional human. Edgar is currently single and looking for a younger housekeeper companion ☺. He can be contacted at Edgar@EdgarLOwen.com.

www.ingramcontent.com/pod-product-compliance
Lightning Source LLC
Chambersburg PA
CBHW080652190526
45169CB00006B/2080